S.O.S

Spotlight on Standards!

Interactive Science Content Reader

California Science

Printed in the United States of America

ISBN-13: 978-0-15-365365-0
ISBN-10: 0-15-365365-5

1 2 3 4 5 6 7 8 9 10 030 16 15 14 13 12 11 10 09 08 07

Standard Set 1 Physical Science

Unit 1 Elements and Compounds

1 Elements and their combinations account for all the varied types of matter in the world.

Standard Set 2 Life Science

Unit 2 Structures of Living Things

2 Plants and animals have structures for respiration, digestion, waste disposal, and transport of materials.

Standard Set 3 Earth Science

Unit 3 The Water Cycle

3 Water on Earth moves between the oceans and land through the processes of evaporation and condensation.

Standard Set 4 Earth Science

Unit 4 Weather

4 Energy from the Sun heats Earth unevenly, causing air movements that result in changing weather patterns.

Standard Set 5 Earth Science

Unit 5 The Solar System

5 The solar system consists of planets and other bodies that orbit the sun in predictable paths.

Elements and Compounds

In this unit, you will learn what atoms, elements, and metals are. You will also learn about the properties of some common substances. What do you know about these topics? What questions do you have?

Thinking Ahead

What is an atom? Write what you think.

Draw an object that is commonly made of metal.

If you had to identify an unknown substance, what properties could you look for?

Suppose you combined vinegar and baking soda. Draw pictures to show what you might see.

+ =

Write a question you have about elements and compounds.

Recording What You Learn

On this page, record what you learn as you read the unit.

Lesson 1

What is the smallest unit of an element?

____

Lesson 2

Draw an object made of metal. Write two of its physical properties.

____

____

Lesson 3

What is the most common element on Earth? What are some of its physical properties?

____

____

____

Lesson 4

Name three properties that can help identify a substance. They can be chemical or physical properties.

____

Lesson 5

If an electric charge goes through water, what happens? What kind of reaction is this?

____

California Standards in This Lesson

I.b *Students know* all matter is made of atoms, which may combine to form molecules.

I.d *Students know* that each element is made of one kind of atom and that the elements are organized in the periodic table by their chemical properties.

I.f *Students know* differences in chemical and physical properties of substances are used to separate mixtures and identify compounds.

Vocabulary Activity

Physical Properties

A *physical property* is any property of matter that can be observed by your senses. The properties don't change when they are measured. Name as many physical properties as you can think of.

________________ ________________

________________ ________________

________________ ________________

________________ ________________

Lesson 1

What Are Atoms and Elements?

VOCABULARY

element
atom
periodic table
compound
molecule
physical property
mixture

These gold bars have only one kind of **element** in them.

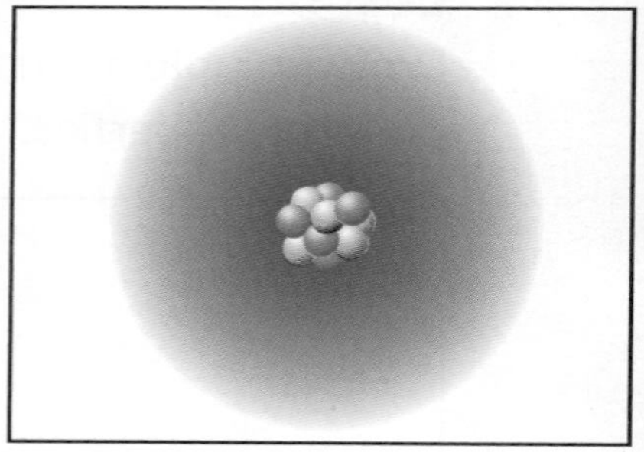

An **atom** is a very small unit of matter.

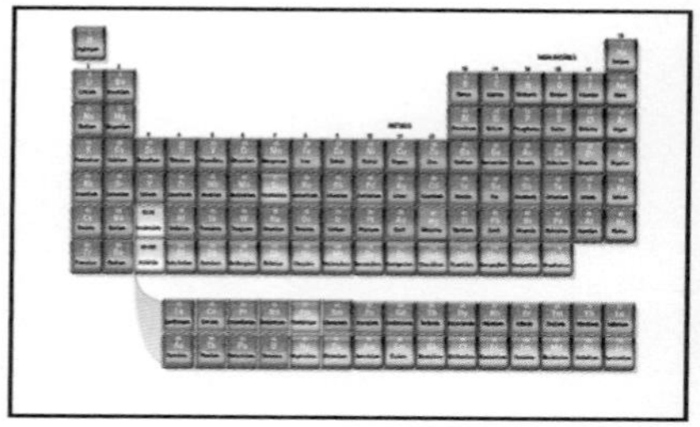

Scientists use the **periodic table**. It tells them a lot about the elements.

Vinegar
Baking Soda

These **compounds** are made from different elements.

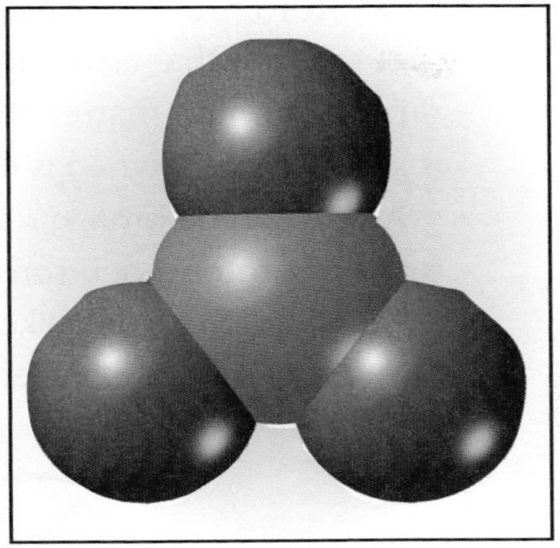

Atoms of different elements join to make a **molecule**.

The shape of this lizard is a **physical property**.

This salad is made from a **mixture** of different fruits.

Hands-On Activity
Observe

1. Observe three common objects in the classroom. Describe each object on the lines below. Write the physical properties. Then read each description aloud to a partner. Have your partner try to identify the object you describe. Then switch roles.

   Object 1: ______________________________

   ____

   Object 2: ______________________________

   ____

   Object 3: ______________________________

   ____

2. What are your partner's objects?

   ____

3. How many of your partner's objects did you guess correctly?

   ____

✓Concept Check

1. The **Main Idea** on these two pages is An atom is the smallest unit of an element. **Details** tell more about the main idea. Find details about what an atom is. Underline two of them.

2. What is the smallest part of an element that has all of the element's properties?

3. What are atoms made of?

4. Complete the chart. Put a check in the correct column.

	Has Properties of an Element	Does Not Have Properties of an Element
Atom		
Proton		
Neutron		
Electron		
Oxygen		

Atoms and Elements

An **atom** is the smallest unit of an element. It is the smallest part that has that element's *properties*. An element is the simplest form of matter. It has only one kind of atom. Oxygen is an element. Oxygen is made only of oxygen atoms.

Atoms can be broken down into smaller parts. Atoms are made of protons, neutrons, and electrons. All protons are alike. All neutrons are alike. All electrons are alike. They do not have properties of any element.

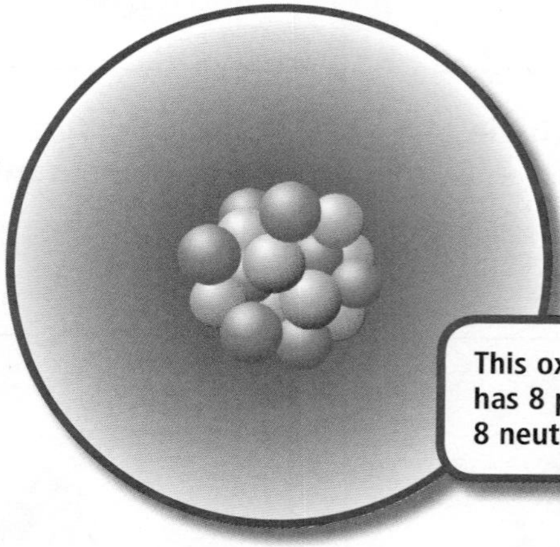

This oxygen atom has 8 protons and 8 neutrons.

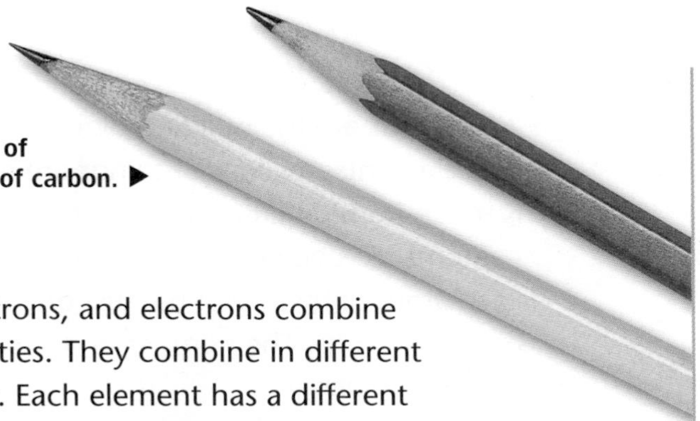

The lead of a pencil is made of graphite. This is a soft form of carbon. ▶

The ways protons, neutrons, and electrons combine form an element's properties. They combine in different amounts in each element. Each element has a different number of protons. The number of protons tells scientists which element it is.

Atoms have the properties of the element they are part of. Carbon atoms have the properties of carbon.

There are about 100 elements. But only a few are found in their pure states. Scientists give each element a short name. They use one to three letters to name each element. Carbon is *C*. Sodium is *Na*.

All living things have carbon in them. ▶

◀ A diamond is made of almost pure carbon.

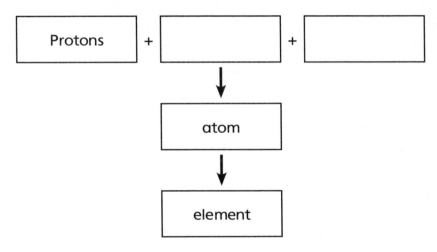

✓Concept Check

1. What makes an element different from other elements?

____

2. What gives an element its properties?

____

3. Underline the sentences that tell how elements are named.

4. Show what an element is made of. Finish the chart.

Protons	+	________	+	________

↓

atom

↓

element

✓Concept Check

1. The **Main Idea** on these two pages is Everything in nature is made of elements. **Details** tell more about the main idea. Find details about elements. Underline two of them.

2. The table below shows symbols for elements. Look at the periodic table. Fill in the name of each element. Then write how many protons each has. The first is done for you.

Symbol	Name	Number of Protons
Na	Sodium	11
H		
O		
C		

The Periodic Table

Elements are the building blocks of matter. From only about 100 elements, millions of things are made. Everything on Earth is made from different groups of elements.

Atoms of each element have a certain number of protons. Each atom of hydrogen has just one proton. Each atom of iron has 26 protons. The number of protons in an atom is called its *atomic number*.

Scientists use a chart called the **periodic table**. In this chart, the atomic numbers of all elements are shown. Each element has its own box in the table. In that box is the atomic number of the element. Also in the box are the element's name and symbol.

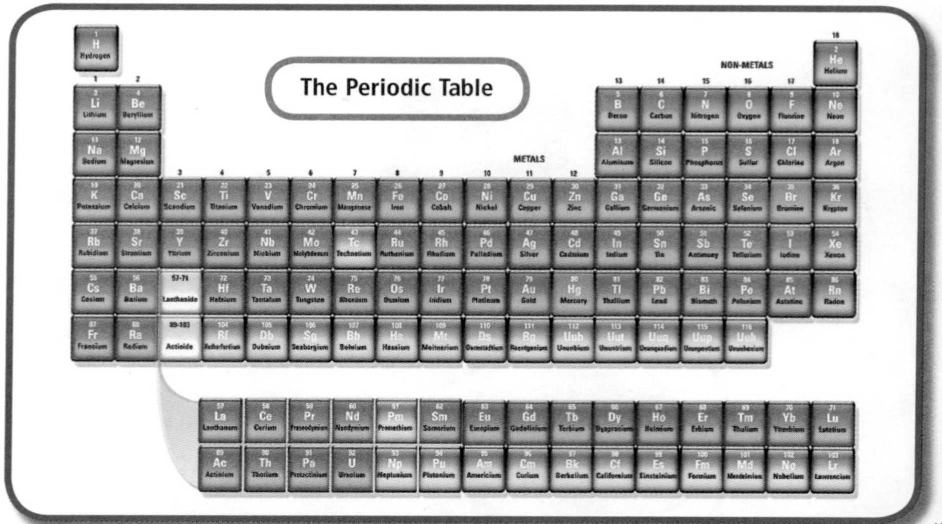
The Periodic Table
NON-METALS
METALS

Molecules and Compounds

Most elements are seen as parts of compounds. A **compound** is a substance made of two or more elements. The elements have been joined chemically. A *formula* tells what is in a compound. It shows how many atoms of each element are in the compound.

H_2O is the formula for the compound *water*. Water is made of hydrogen and oxygen atoms. Both of these are elements.

The smallest piece of the compound water is a water molecule. A **molecule** is the smallest unit of a compound.

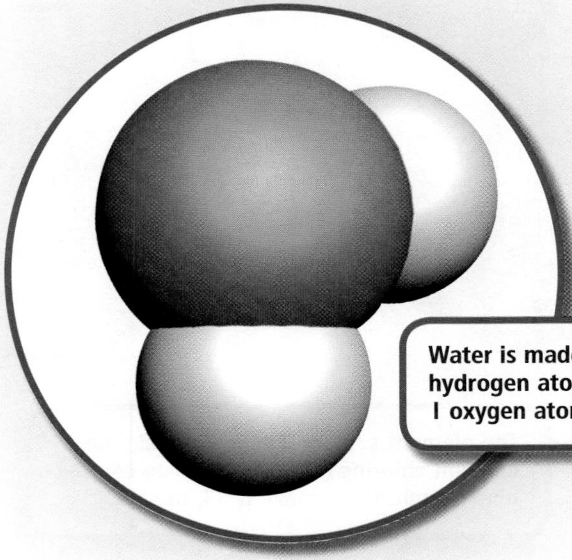

Water is made of 2 hydrogen atoms and 1 oxygen atom.

✓Concept Check

1. What elements make up a water molecule?

2. Circle the sentence that describes a compound.

3. Look at pages 4 and 5. Then finish the chart about atoms and molecules. Put check marks in the correct columns.

	Atom	Molecule
Smallest unit of an element		
Smallest unit of a compound		
Made of one element		
Made from two or more elements		

✓Concept Check

1. The **Main Idea** on these two pages is Physical properties describe a substance. **Details** tell more about the main idea. Find details about physical properties. Underline two of them.

2. Give two examples of a physical property you can see or feel.

____

3. Give two examples of a physical property you can NOT see.

____

4. What is one physical property of nitrogen?

____

Physical Properties

Physical properties describe a substance. They tell what a substance is like by itself. Water is wet. Wood is hard. These are physical properties.

Some physical properites cannot be seen directly. Temperature is a physical property. But you cannot see it. Mass is a physical property. But you cannot see it. Both mass and temperature have to be measured.

SODIUM
Chlorine

The element sodium (Na) is a metal. The element chlorine (Cl) is a gas. They combine chemically to make NaCl, or salt.

A **mixture** is a combination of two or more substances. In a mixture, the substances themselves are not changed. Each part of the mixture could be separated out again.

Most materials found in nature are mixtures. Sand and air are both mixtures.

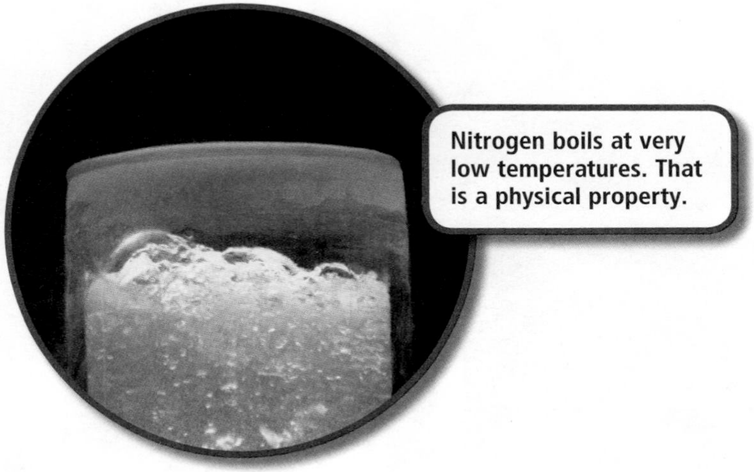
Nitrogen boils at very low temperatures. That is a physical property.

Lesson Review

Complete this Main Idea statement.

1. ____________ are the building blocks of matter.

Complete these Detail statements.

2 Scientists use the ____________ ____________ to learn about elements.

3. A ____________ is the smallest unit in which an element can exist.

4. In a ____________ physical, but not chemical, changes take place.

California Standards in This Lesson

1.c *Students know* metals have properties in common, such as high electrical and thermal conductivity. Some metals, such as aluminum (Al), iron (Fe), nickel (Ni), copper (Cu), silver (Ag), and gold (Au), are pure elements; others, such as steel and brass, are composed of a combination of elemental metals.

1.e *Students know* scientists have developed instruments that can create discrete images of atoms and molecules that show that the atoms and molecules often occur in well-ordered arrays.

Vocabulary Activity

There are three vocabulary terms that feature the word *metal*. The prefix *non-* means "not." The suffix *–oid* means "like."

1. Use the information above to define two of the vocabulary words.

Word	Meaning
metalloid	
nonmetal	

2. Read the descriptions for the words malleable and alloy. Which word is an adjective? ______________

3. Which word is a noun? ______________

Lesson 2

What Are Metals?

VOCABULARY

metal
nonmetal
malleable
alloy
metalloid

Metal is used to build things that need to last a long time.

This lump of coal is made of carbon, a **nonmetal**.

The aluminum in this can is very **malleable**.

This statue is made of bronze. Bronze is a metal **alloy**.

Computer chips are made from silicon, a **metalloid**.

Hands-On Activity
Draw Conclusions

1. Fill three glasses with water of the same temperature. Get a sample of a metal, an alloy (bronze or tin), and a nonmetal. Ask an adult to help you place each sample in a very warm place, such as on a radiator. Get three thermometers that all show the same temperature.

2. After five minutes, place each sample in its own glass of water. Wait one minute.

   Predict which cup of water will be warmest.

   ____

3. Place one thermometer in each glass and measure the water temperature.

   Which cup had the highest temperature?

   ____

3. What made the water hotter?

   ____

4. Draw conclusions about which substance conducts heat best.

   ____

✓Concept Check

1. The **Main Idea** on these two pages is Metals and nonmetals have different properties. **Details** tell more about the main idea. Find details about the properties of metals and nonmetals. Underline one of each.

2. List three properties of metals.

____

3. A circle graph can help you show percentages. Color the circle graph below. Use blue for metals and red for nonmetals. Then label each section.

____

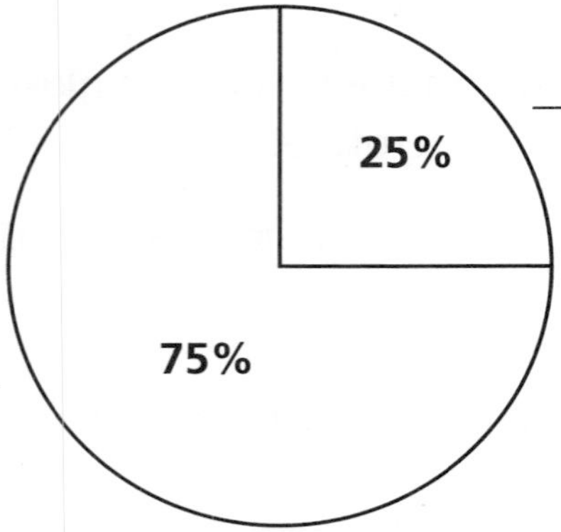
25%
75%

____

Metals and Nonmetals

A **metal** is a substance that easily conducts heat and electricity. This means that metals heat up quickly. Electricity passes through them easily.

About 75% of all elements are metals.

Copper is a metal.

Copper is used for wires. This is because it is a good conductor of electricity. ▼

Metals are **malleable**. They can be hammered or rolled into thin sheets. They can be made into different shapes.

Carbon is a **nonmetal**. Nonmetals do not have the properties of metals. Nonmetals do not conduct electricity well. They do not conduct heat well.

Mercury is a metal. It remains a liquid at room temperature.

✓Concept Check

1. Name two properties of nonmetals.

____

2. How are metals different from nonmetals?

____

3. Metal is malleable. Circle the sentences that tell what that means.

4. Name a metal that remains liquid at room temperature.

____

5. Give an example of a nonmetal.

____

✓Concept Check

1. The **Main Idea** on this page is Metals have properties. **Details** tell more about the main idea. Underline two details about properties of metals.

2. The picture shows an element on the periodic table.

13 **Al** Aluminum

How is this element like titanium?

___

3. These elements are near each other on the periodic table. Why do you suppose they are near each other?

15	16	17	18
			2 He Helium
7 N Nitrogen	8 O Oxygen	9 F Fluorine	10 Ne Neon
15 P Phosphorus	16 S Sulfur	17 Cl Chlorine	18 Ar Argon
33 As Arsenic	34 Se Selenium	35 Br Bromine	36 Kr Krypton

___

Properties of Metals

Elements in columns in the periodic table are alike. For example, some metals conduct heat better than others.

Some metals are much less dense than others. Aluminum and titanium are two such metals. They are very light. They are used to make airplanes and other things that need to be light.

Iron is a good conductor of heat. It is often used to make cooking pans.

Seeing Metal Atoms

Atoms are very small! Scientists must magnify atoms in order to see them. They use special microscopes.

Most microscopes magnify things about 1,000 times. An *electron microscope* can magnify more than 50,000 times. With an electron microscope, scientists can see very tiny objects, like cells. Another powerful microscope is called a *scanning tunneling microscope*, or STM. This microscope lets scientists look at atoms a different way. It makes images based on how electrons move. The atoms in a solid are arranged in set patterns. This is called an *array*.

Scientists can now use an STM to move single atoms to the surface of a metal. They hope someday to make new materials this way.

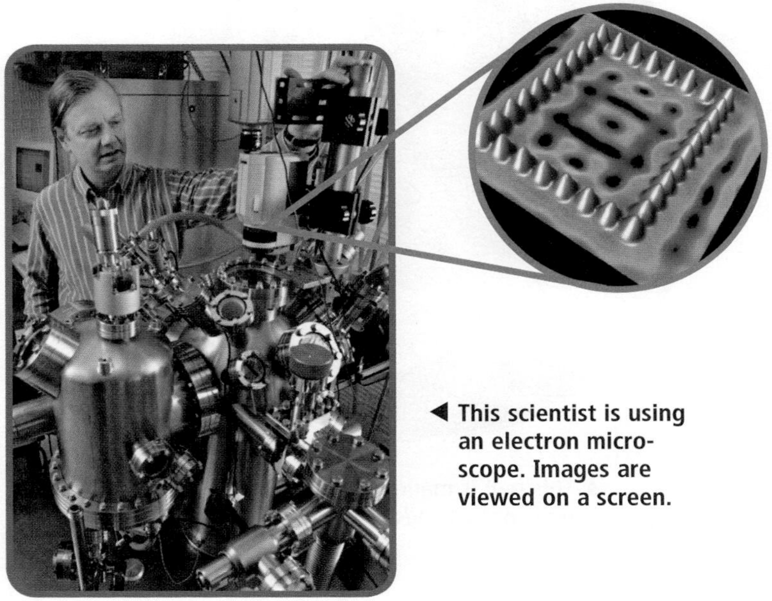

◀ **This scientist is using an electron microscope. Images are viewed on a screen.**

✓Concept Check

1. Name two tools scientists use to see atoms.

____

2. How are atoms arranged in solids? Underline the sentences that tell you.

3. What does STM stand for?

   S _______________

   T _______________

   M _______________

4. What does an STM do that other microscopes do not?

____

✓Concept Check

1. The **Main Idea** on these two pages is Alloys and metalloids share some physical properties with metals. **Details** tell more about the main idea. Find details about the physical properties of alloys and metalloids. Underline two of them.

2. Finish the chart to show which elements are in different alloys.

Element	+	Element	=	Alloy
carbon	+		=	steel
copper	+	tin	=	

3. What type of element is silicon?

____

Alloys and Metalloids

Steel is an **alloy**. An alloy is a solid solution. One way to make an alloy is to combine metals. Another way is to combine a metal and a nonmetal. An alloy has properties different from the materials in it.

Iron is the main metal in many steel alloys. But iron is not very hard. It becomes harder when carbon is added to make steel.

Bronze is an alloy made from mixing copper and tin. Bronze is harder and stronger than copper. Many large statues are made of bronze. They can stay outside for a long time and not *corrode*, or break down.

▲ **This bell is made of brass. It does not corrode easily.**

Metalloids are elements. They have some of the properties of metals. They have some of the properties of nonmetals. Silicon is a metalloid.

Silicon is less malleable than metal. It conducts electricity, but not as well as copper. The flow of electricity can be controlled in silicon. Metalloids like this are called *semiconductors*. Semiconductors are used in computers and CD players.

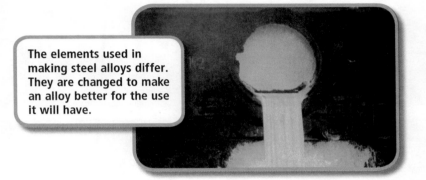

The elements used in making steel alloys differ. They are changed to make an alloy better for the use it will have.

Lesson Review

Complete this Main Idea statement.

1. ____________ is a substance that easily conducts electricity and heat.

Complete these Detail statements.

2. Metals are ____________ They can be hammered or rolled into shapes.
3. An ____________ is made from different metals, or a metal and a nonmetal.
4. A ____________ has properties of both metals and nonmetals.

California Standards in This Lesson

1.g *Students know* properties of solid, liquid, and gaseous substances, such as sugar ($C_6H_{12}O_6$), water (H_2O), helium (He), oxygen (O_2), nitrogen (N_2), and carbon dioxide (CO_2).

1.h *Students know* living organisms and most materials are composed of just a few elements.

Vocabulary Activity

States of Matter

Water exists in each of these states of matter.

1. Write the word for water in each of the states below.

State	Name When Water
Gas	Vapor
Liquid	
Solid	

2. Tell the state of matter for each of the examples. Use each vocabulary word once.

Object	State
Wood	
Juice	
Air	

Lesson 3

What Are the Properties of Some Common Substances?

VOCABULARY
gas
liquid
solid

These balloons are filled with helium **gas**.

Skim milk is a **liquid** that is healthful to drink.

The sand in this sand castle is a **solid**.

Hands-On Activity
Predict

1. Fill a plastic water bottle with water. Weigh the filled bottle. Write how much the bottle weighs. Draw what it looks like.

____

2. Place the full water bottle in the freezer. Predict how much the bottle will weigh when it is frozen. Draw what you predict it will look like when it is frozen.

____

3. Leave the water bottle in the freezer overnight. The next day, weigh the bottle. Write how much it weighs. Draw what it looks like.

____

4. What conclusion can you draw?

____

✓Concept Check

1. The **Main Idea** on these two pages is <u>There are many gases.</u> **Details** tell more about the main idea. Find details about which gases are common. Underline two of them.

2. Name two physical properties of gases.

____

3. What gas do you need to have a fire?

____

4. Look at the chart called "Gases in the Atmosphere." Complete the chart below to show the gases in our atmosphere. List the gases in order from least to most common.

Gas	Percentage
Other gases or Carbon dioxide	0.035%
	21%
	78%

Common Gases

Air is invisible. Air is not an element or a compound. Air has no chemical formula. It is a mixture of gases.

A **gas** is a state of matter. In a gas, matter has no definite shape. It has no definite *volume*, or amount, either.

Air is about 78% nitrogen (N_2). Air also has about 21% oxygen (O_2) in it. Oxygen makes it possible for other substances to burn. They burn when they mix with oxygen.

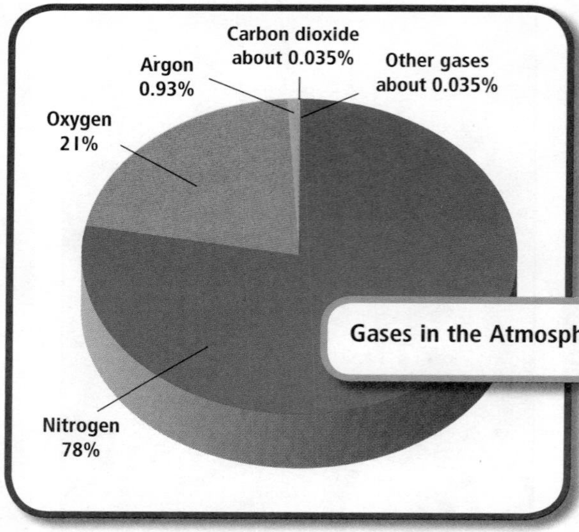
Carbon dioxide about 0.035%
Argon 0.93%
Other gases about 0.035%
Oxygen 21%
Nitrogen 78%

Gases in the Atmosphere

Another form of oxygen is ozone (O_3). There is ozone in Earth's upper atmosphere. Ozone helps keep Earth at a constant temperature.

Hydrogen is another gas. It is the most common element in the universe.

Helium is a gas that is lighter than air. It is used in weather balloons and blimps.

Another common gas is carbon dioxide (CO_2). When you breathe out, you *exhale* mostly carbon dioxide.

This balloon is filled with helium. Helium weighs less than air, so the balloon floats. ▼

✓Concept Check

1. Why doesn't air have a chemical formula?

2. Which gas helps keep Earth at a constant temperature?

3. Which element is more common: helium or hydrogen?

4. This boy breathes out as he blows bubbles.

What gas is he breathing out?

✓Concept Check

1. The **Main Idea** on these two pages is Liquids have their own properties. **Details** tell more about the main idea. Find details about the properties of liquids. Underline two of them.

2. Why is water unlike any other compound?

____

____

3. Check the spaces that show the correct physical properties. Look at pages 20–23 for help.

	Gas	Liquid
No definite shape		
No definite volume		
State of matter		
Commonly found on Earth		

Common Liquids

A **liquid** is a state of matter. In this state, matter has a definite volume. But it has no definite shape.

Water is the most common liquid on earth. The oceans cover more than 70% of the earth. Ocean water is salt water.

Water is the only compound on Earth that can be found in all three states. Water can be a liquid. It can be a gas (water vapor). Or it can be a solid (ice).

These monkeys in a hot spring have ice on their heads. Water is seen in all three states here. ▼

When most liquids are cooled, their particles get closer together. They *contract*, or take up less space. Water contracts until it reaches about 4 degrees Celsius (39° F). As it cools further, it *expands*, or takes up more space.

As water molecules change into ice, they form rings. The rings take up more space. That is why ice takes up more space than water.

Mercury is the only metallic element that is liquid at normal temperatures. It was used in thermometers. But mercury is poisonous, so it is not used any more. Alcohol is used now.

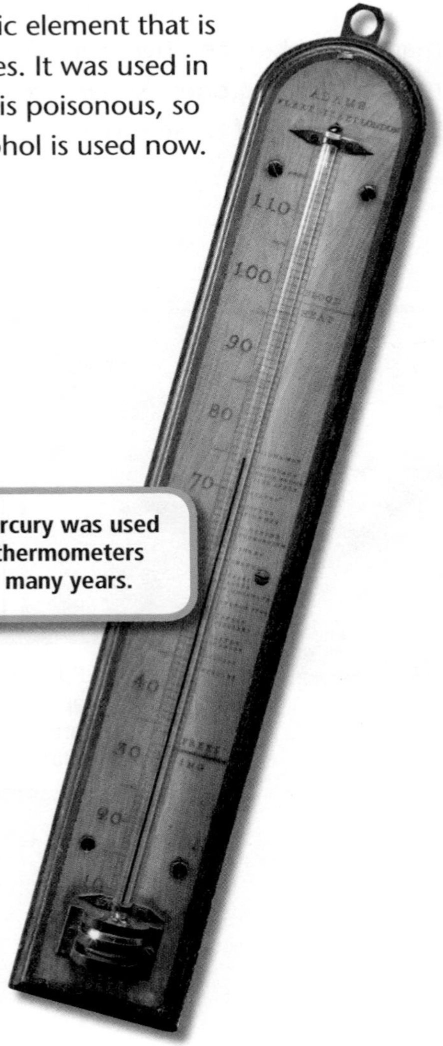

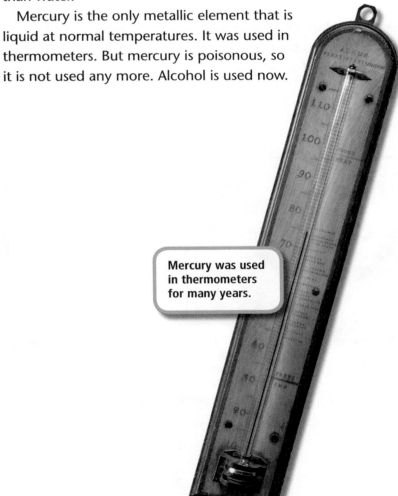

Mercury was used in thermometers for many years.

✓Concept Check

1. What are some unusual properties of water?

2. Look at the thermometer. Draw an arrow pointing to the temperature at which the water in the cup will begin expanding.

3. What makes water expand as it cools?

✓Concept Check

1. The **Main Idea** on these two pages is Solids are common and have their own properties. **Details** tell more about the main idea. Find details about the properties of solids, and how they are commonly found. Underline two of them.

2. What is the sixth most common element?

____

3. Name three things made of carbon.

____

4. Check the spaces that show the right physical properties.

	Gas	Liquid	Solid
Definite shape			
Definite volume			
State of matter			
Commonly found on Earth			
Carbon is one			

Common Solids

The element carbon is a solid. A **solid** is a state of matter. It has a definite shape and a definite volume.

Carbon is the sixth most common element in the universe. There are millions of compounds that have carbon in them. Charcoal, the "lead" in a pencil, and diamonds are all carbon.

Coal, petroleum, and natural gas are compounds. They are made of carbon and hydrogen. When oxygen is added to carbon and hydrogen, sugars are made. Corn syrup and table sugar are two of these sugars.

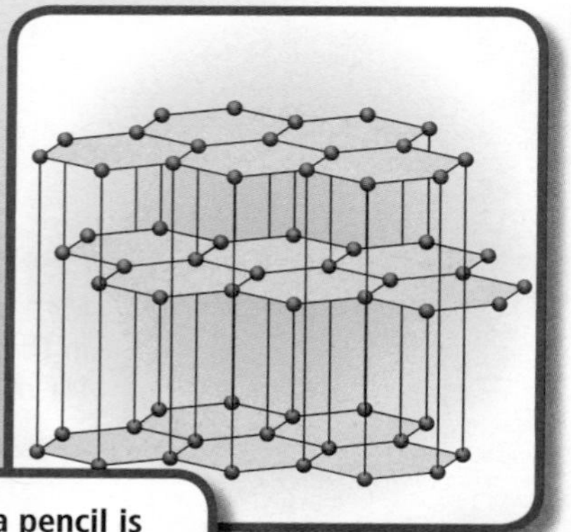

The "lead" in a pencil is called graphite. In graphite, the carbon atoms are joined in flat layers.

Copper and nickel come from Earth. They are used in making coins. Stones for fireplaces also come from Earth. Aluminum for cans and silicon for glasses come from Earth's crust.

Gemstones, like rubies and diamonds, are mined from under the ground. Metals for cars and ovens come from materials found in Earth's crust.

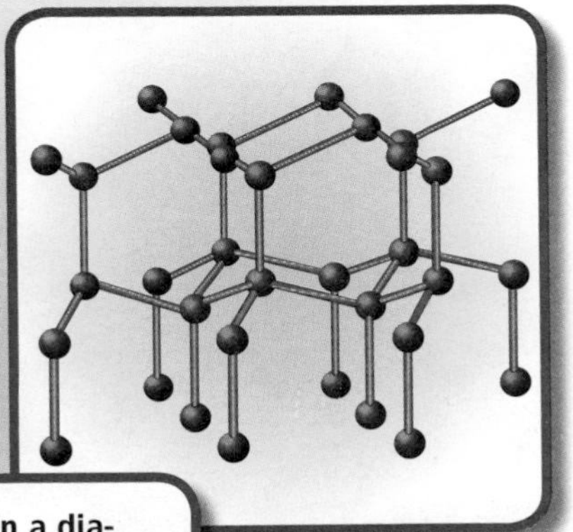

The atoms in a diamond are arranged in a way that makes it very hard.

✓ Concept Check

1. Where do most of the objects you use originally come from?

____

2. Name three items that come from materials found inside Earth.

____

____

3. The diagrams show the way atoms are arranged in graphite and diamonds. Label each.

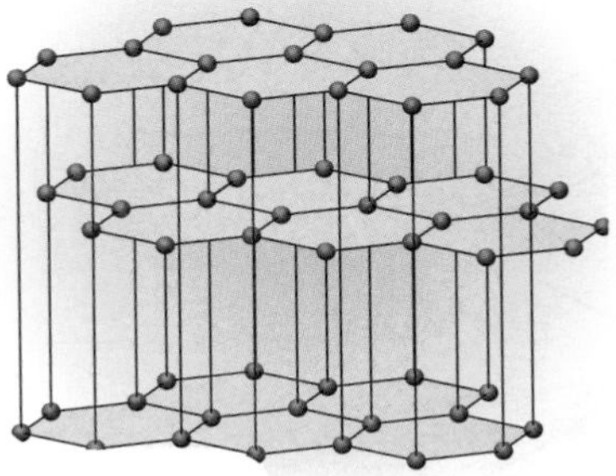

_______________ _______________

✓ Concept Check

1. The **Main Idea** on these two pages is A few elements make up most life forms. **Details** tell more about the main idea. Find details about elements. Underline two of them.

2. Which element makes up most of the human body?

____

3. What element is the basic building block of the molecules of life?

____

4. Fill in the missing labels on the chart.

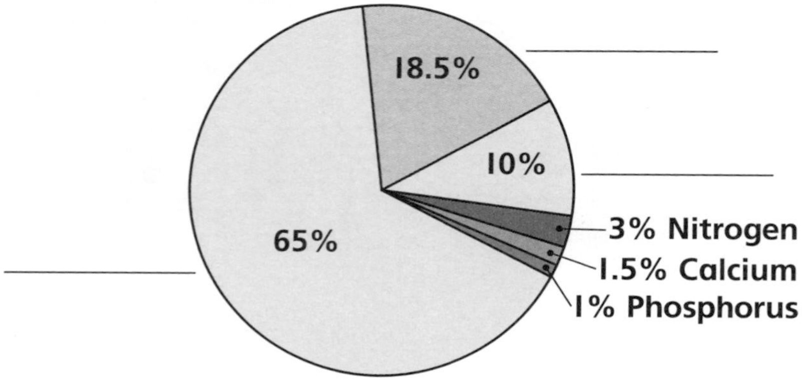

18.5%
10%
65%
3% Nitrogen
1.5% Calcium
1% Phosphorus

Molecules of Life

Without carbon, Earth would be lifeless. Even tiny bacteria are 50% carbon. Carbon is the basic building block of life.

About 98.8% of the human body is made from 6 elements. Oxygen, at 65%, makes up the largest amount. Carbon is second, at 18.5%. Hydrogen, nitrogen, calcium, and phosphorus are the other four.

Each element is needed for good health.

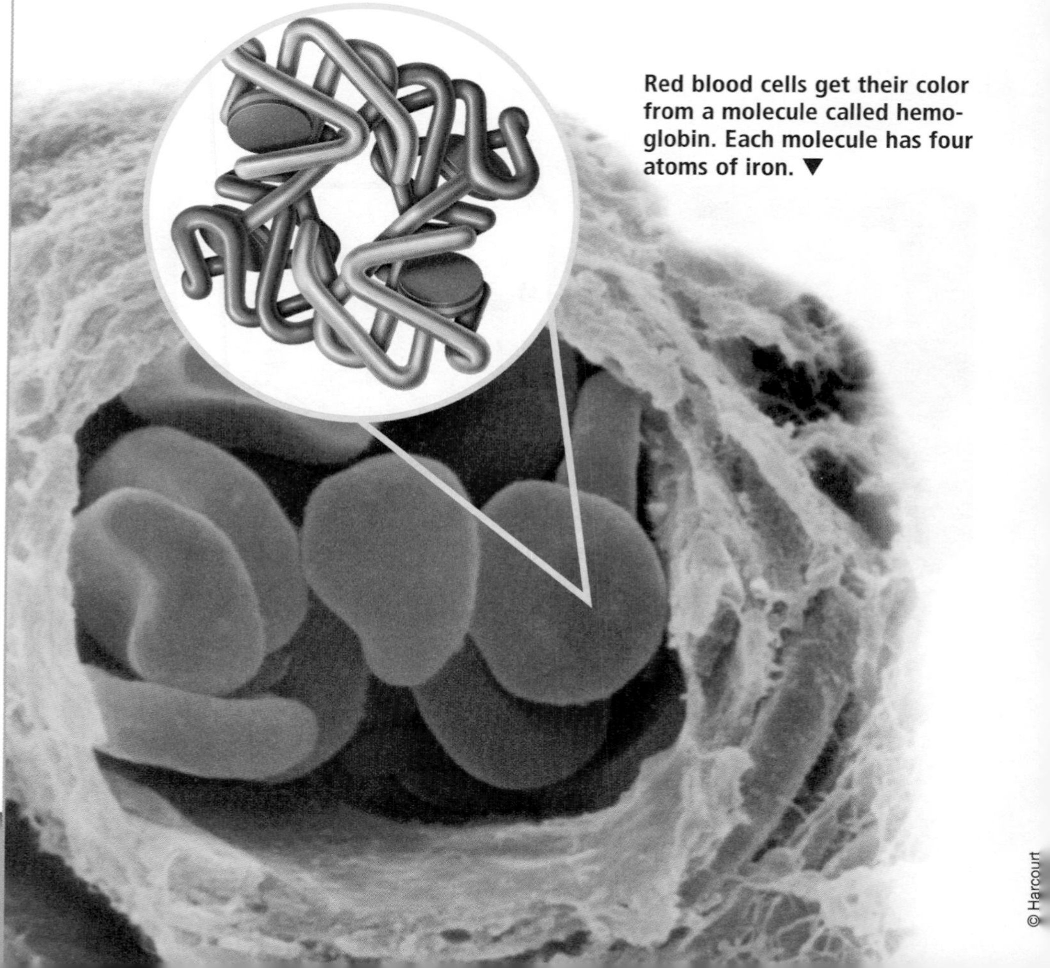

Red blood cells get their color from a molecule called hemoglobin. Each molecule has four atoms of iron. ▼

DNA is in every cell of the body. It makes sure that when new cells are made, they are exactly like the old ones. But it is made from just a few elements.

The DNA molecule looks like a twisted ladder. Each colored ball in this picture of DNA shows a different element.

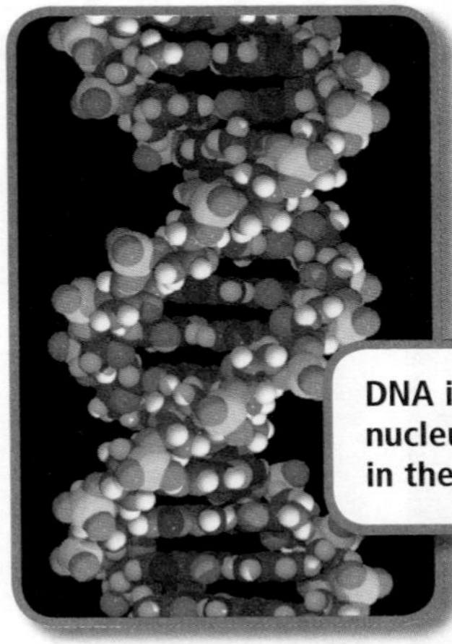

DNA is found in the nucleus of every cell in the body.

Lesson Review

Complete this Main Idea statement.

1. All ____________ comes in three states: solid, liquid, or gas.

Complete these Detail statements.

2. A ____________ has no definite shape and no definite volume.
3. A ____________ has a definite volume but no definite shape.
4. A ____________ has a definite shape and a definite volume.

California Standards in This Lesson

1.f *Students know* differences in chemical and physical properties of substances are used to separate mixtures and identify compounds.

Vocabulary Activity

You learned about physical properties. Now you are learning about chemical properties. Physical reactions and chemical reactions are different types of reactions. An acid is the opposite of a base.

1. Name another pair of opposites.

____

2. Adding the suffix *–ic* turns *base* and *acid* into adjectives. Read the sentences about acids and bases on page 28 and 29. Now, rewrite them using the adjectives *acidic* and *basic*.

____

____

Lesson 4

How Are Chemical and Physical Properties Used?

VOCABULARY
chemical property
acid
base

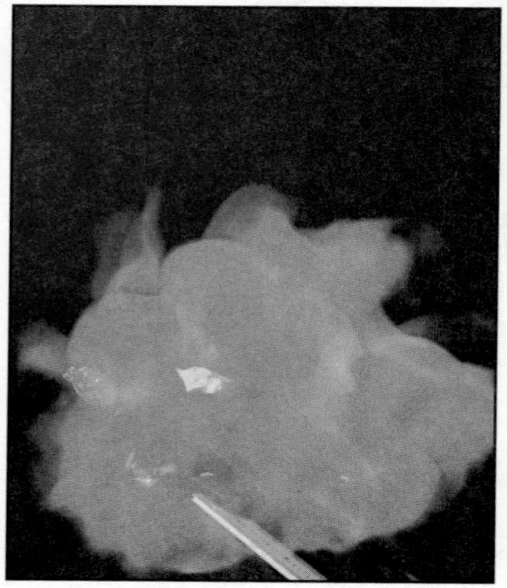

Flammability, or the ability to catch on fire, is one kind of **chemical property**.

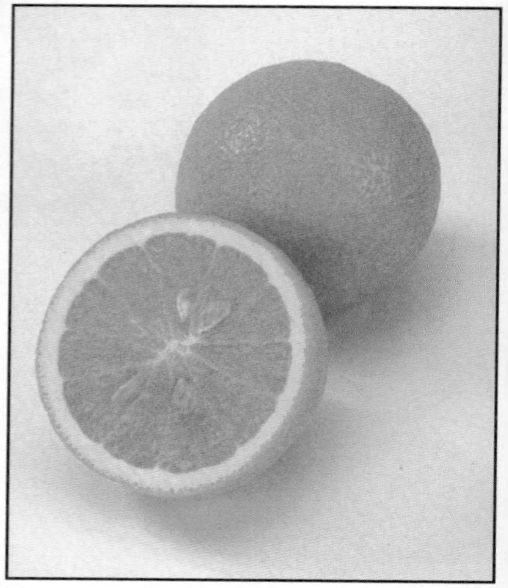

The juice in these oranges is a weak **acid**.

This soap has the properties of a weak **base**.

Hands-On Activity
Collect Data

1. Get some litmus paper, juices, sodas, vinegar, and dish soap. Using a dropper, place one drop of each liquid on a different piece of litmus paper. Be sure to rinse the dropper carefully before putting it into a new liquid!

2. Record the kind of liquid and the pH. Write whether each liquid is an acid, a base, or neutral.

___

___

Kind of Liquid	pH	Acid, Base, or Neutral?

✓Concept Check

1. The **Main Idea** on these two pages is Physical and chemical properties can be used to separate mixtures. **Details** tell more about the main idea. Find details about how to separate mixtures chemically. Underline one of them.

2. Give an example of when it would be hard to separate a mixture.

____

3. Write the steps you would take to separate a mixture of rocks, dirt, iron, and salt.

   1. ______________________________
   2. ______________________________
   3. ______________________________
   4. ______________________________

Separating Mixtures

Sometimes it is easy to separate the parts of a mixture. At other times, it is not so easy. It would be easy to separate raisins from flakes in cereal. It would not be easy to separate flour and sugar mixed together.

Physical properties can help you separate mixtures. The picture shows how to separate a mixture of rocks, dirt, iron and salt.

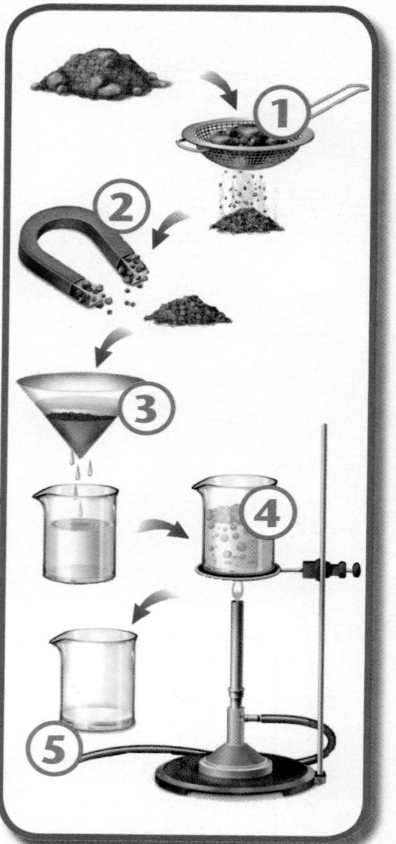

1
2
3
4
5

To separate a mixture of rocks, dirt, iron and salt: 1. Screen out the rocks. 2. Use a magnet to take away bits of iron. 3. Pour water through a filter to remove dirt. 4. Boil the water. 5. Only salt is left behind.

Chemical properties can also be used to separate mixtures. A chemical property describes a *reaction*. It shows how substances react when they combine to make new substances.

Most metals in nature are found in minerals. These minerals are called *ores*. Ores are compounds.

The metal must be removed from the ore before it can be used. Chemicals are used to do this. Then the metal can be used.

In nature, copper is usually found combined with oxygen or sulfur.

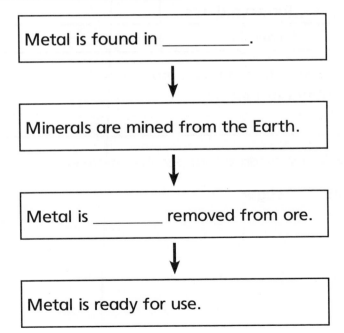

✓Concept Check

1. How are physical properties used to separate mixtures?

____

____

2. In what form are most metals found in nature?

____

3. Finish the chart. Show the steps you need to take before metal can be used.

Metal is found in __________.
↓
Minerals are mined from the Earth.
↓
Metal is ________ removed from ore.
↓
Metal is ready for use.

✓Concept Check

1. The **Main Idea** on these two pages is Elements and compounds can be identified physically and chemically. **Details** tell more about the main idea. Find details about how elements and compounds can be identified. Underline two of them.

2. Suppose you did a litmus test. The results are shown below. Write whether the tests showed an acid or a base.

Results	Acid or Base?
Turned red	
Turned blue	

3. What are three physical properties that help identify substances?

____

4. Circle a substance that dissolves more easily in water.

salt **sugar**

Identifying Elements and Compounds

Chemical properties help identify elements and compounds. You can find out if a substance is an **acid** or a **base**. *Litmus paper* is used. It has chemicals in it that change colors. An **acid** turns blue litmus paper red. A **base** turns red litmus paper blue.

Some acids and bases are stronger than others. Scientists rank acids and bases on a scale. This scale, called the *pH scale*, has numbers from 0 to 14. An acid has a pH of less than 7. A base has a pH of more than 7.

Water and some other substances are *neutral*. They have a pH of 7. They are neither acids nor bases.

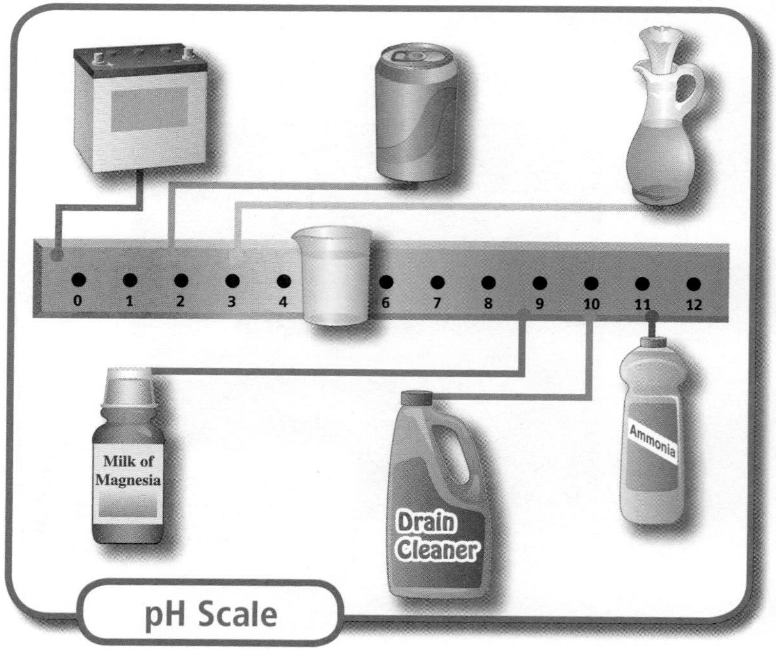
0
1
2
3
4
6
7
8
9
10
11
12
Milk of Magnesia
Drain Cleaner
Ammonia
pH Scale

▲ **Battery acid is strong. It can have a pH of 0. Lemon drink has a pH of about 2. Drain cleaners are stong bases. They have a pH of about 10. Ammonia has a pH of about 11.**

Physical properties help identify substances. Metals can be identified by their melting points. Texture is another physical property. Some things are *brittle*. They break or crush easily. *Solubility* also helps identify elements and compounds. Sugar dissolves easily in water. Salt dissolves less easily.

Chemical properties can also be used to identify substances. But, you have to change the substance itself.

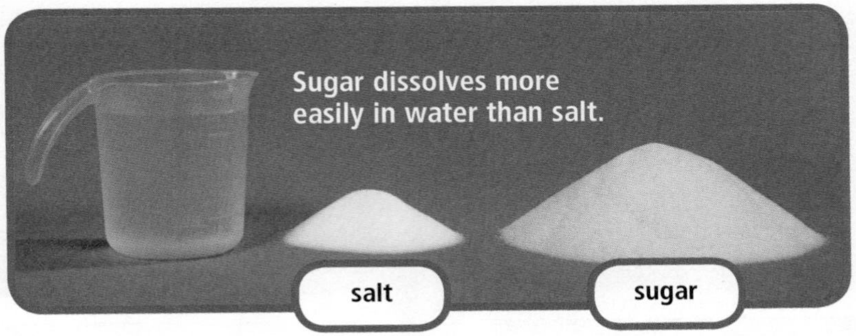
salt
sugar

Sugar dissolves more easily in water than salt.

Lesson Review

Complete this Main Idea statement.

1. Physical and chemical ____________ help identify elements and compounds.

Complete these Detail statements.

2. ____________ ____________ changes color when put into an acid or a base.
3. An ____________ turns blue litmus paper red.
4. A ____________ turns red litmus paper blue.

California Standards in This Lesson

1.a *Students know* that during chemical reactions the atoms in the reactants rearrange to form products with different properties.

1.f *Students know* differences in chemical and physical properties of substances are used to separate mixtures and identify compounds.

Vocabulary Activity

Root Words

The verb *react* means "to act in response." Two of the vocabulary terms have this word in them.

1. Use the table to figure out what the words mean. Use your best guess.

Suffix: meaning	Word: meaning
-ant: one that performs a specific action	*Reactant:*____________
-ion: the result of an action or process	*Chemical reaction:* ____________

2. Write a sentence using the word *product.*

____

Lesson 5

What Are Chemical Reactions?

VOCABULARY
chemical reaction
reactant
product
salt

Rusting is one kind of **chemical reaction**.

Baking Soda
Vinegar

When combined, vinegar and baking soda are **reactants**.

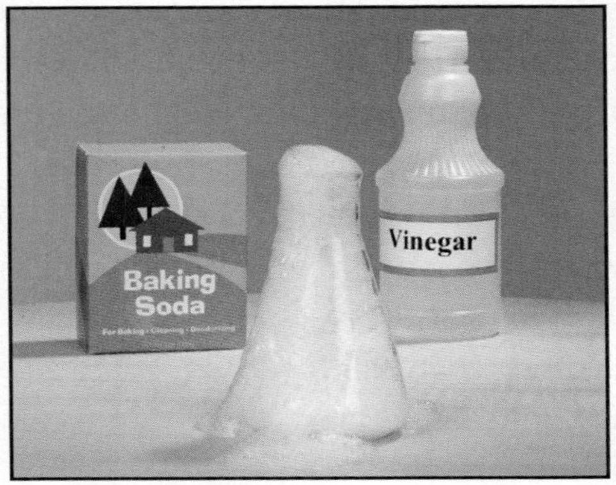
Baking
Soda
Vinegar

Carbon dioxide is a **product**. It comes from combining baking soda and vinegar.

Table **salt** is just one kind of **salt**.

Hands-On Activity
Record Observations

1. Get a cup of vinegar and a spoonful of baking soda. Pour the vinegar into a glass beaker. Carefully add the baking soda.

   Tell two things that happened.

   ____

   ____

   ____

2. What type of reaction is this?

   ____

3. What were the reactants?

   ____

4. What was the product?

   ____

✓Concept Check

1. You **Compare** when you look at how things are alike. You **Contrast** when you look at how things are different. Compare physical and chemical reactions by writing one way they are alike.

____

Contrast physical and chemical reactions by writing one way they are different.

____

____

2. Circle a sentence that describes a physical change.
3. Underline a sentence that describes a chemical change.

Changing Properties

Paper tears. Glass breaks. Sugar dissolves in water. These are physical changes. They are not chemical changes. In a chemical change, the substance cannot be changed back easily.

A **chemical reaction** is a chemical change. If you touch a lit match to wood, the wood burns. When something burns, it combines with oxygen. This produces other gases. You may not see the gases. But you know that something is gone. You see the ashes.

When candles burn, oxygen from the air causes a chemical change.

You can tell when a chemical change takes place. You know because the properties of the materials are different. New substances with different properties have formed.

There are clues that tell you when a chemical change is taking place. When you see bubbles after adding one substance to another, this is a clue. When you see color changes, this is another clue.

	Physical Properties	Chemical Properties
Water 	• colorless • odorless • liquid at room temperature • boils at 100°C • melts at 0°C	• made up of hydrogen and oxygen • reacts with some metals to produce bases
Silver 	• shiny • soft • silver in color • boils at 2163°C • melts at 962°C	• reacts with few substances • does not react with air • reacts with ozone or sulfur to form tarnish
Iron 	• shiny • hard • grayish silver in color • boils at 2861°C • melts at 1538°C	• reacts easily with many substances • reacts with oxygen to form the minerals hematite and magnetite • reacts with oxygen in the presence of water to form rust
Sulfur 	• dull • brittle • yellow • boils at 445°C • melts at 115°C	• reacts with any liquid element • reacts with any solid element except gold and platinum • reacts with oxygen to form sulfur dioxide, a form of air pollution

✓ Concept Check

1. Use the table on page 11 to solve the riddles.

   I react with oxygen to form hematite. I am ____________.

   I boil at 2163°C. I am ____________.

   I am made up of hydrogen and oxygen. I am ____________.

   I react with oxygen to form air pollution. I am ____________.

2. Fill in the chart. Tell whether each is a physical or chemical property.

Property	Chemical or Physical?
Colorless	
Reacts with ozone	
Shiny	
Hard	
Reacts with any liquid element	

Concept Check

1. You **Compare** when you look at how things are alike. You **Contrast** when you look at how things are different. Compare reactants and products by writing one way they are alike.

____

____

Contrast reactants and products by writing one way they are different.

____

____

2. Chemical reactions can be written as formulas. Write the formula below using the words reactant and product. You will use one of the words twice.

___________ + ___________ = ___________

3. Fill in the missing reactants and products.

electricity in water = __________________

sodium + ___________ = sodium chloride

___________ + oxygen = Iron oxide

Reactants and Products

Every chemical reaction begins with one or more **reactants**. A reactant is one of the starting materials in a chemical reaction.

If an electric charge goes through water, bubbles form. The bubbles are hydrogen and oxygen gas. The reactant is water. Hydrogen and oxygen are **products**. They are the result of the chemical reaction.

Sodium and chlorine are elements. Adding water to them makes the sodium and chlorine react. The product of this chemical reaction is sodium chloride. It is table salt.

Another reaction is an iron nail rusting. The iron reacts with oxygen in the air. The iron and oxygen combine. They make the product iron oxide. You know iron oxide as *rust* . Rust has different properties than iron or oxygen.

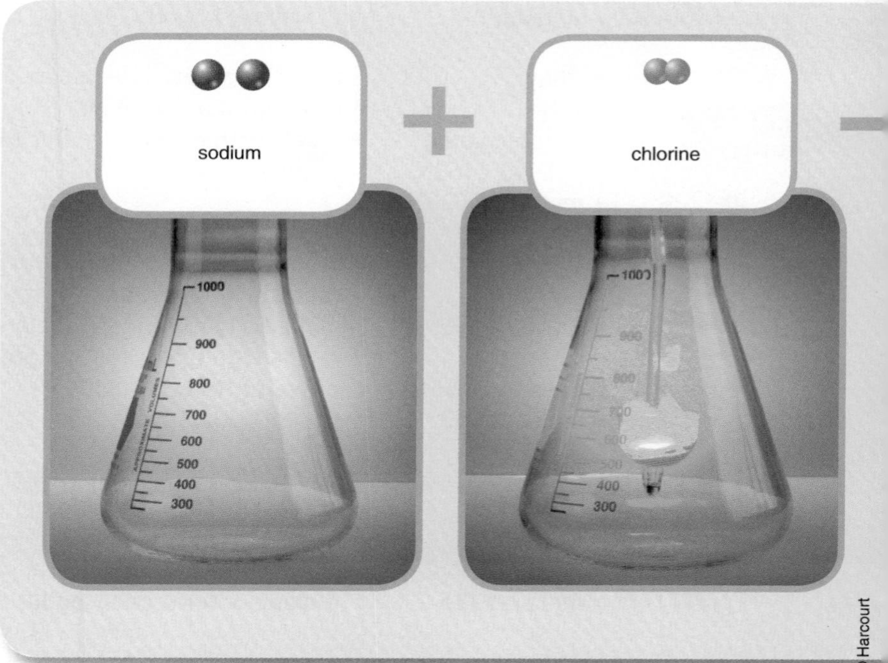
sodium
+
chlorine
1000
900
800
700
600
500
400
300

When silver *tarnishes*, it becomes silver sulfide. One reactant is the element silver. The other reactant is the compound hydrogen sulfide.

Silver reacts with hydrogen sulfide in the air. This makes silver sulfide, or tarnish.

▼ **Sodium reacts with chlorine. Energy is given off as heat and light. Sodium chloride—table salt—is made.**

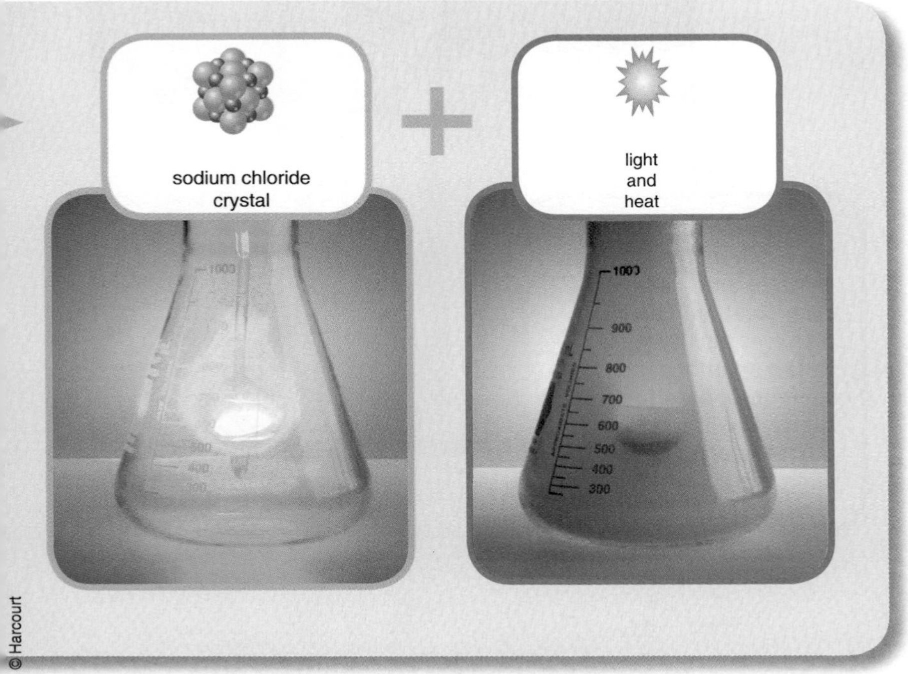
sodium chloride crystal
+
light and heat

Concept Check

1. What kinds of energy are given off when sodium reacts with chlorine?

____

2. Fill in the equation below. Show the chemical reaction that makes silver tarnish. Draw the spoon as it will appear once it becomes silver sulfide.

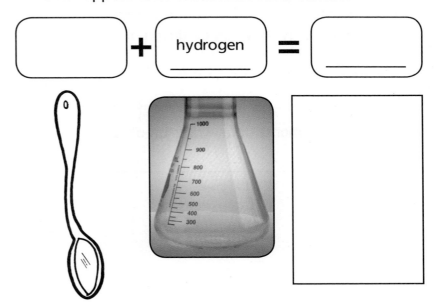

	+	hydrogen ________	=	________

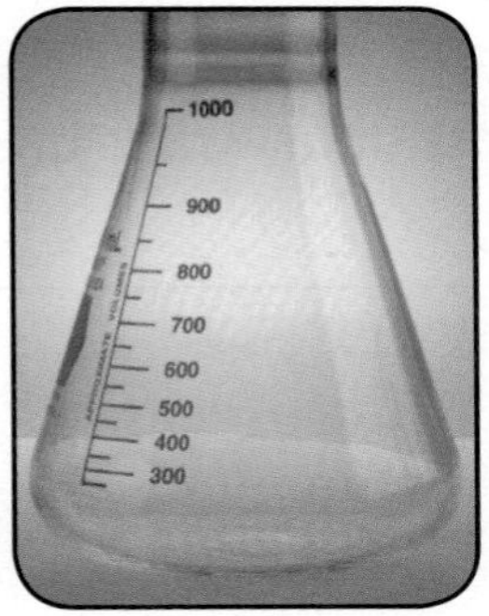

✓Concept Check

1. You **Compare** when you look at how things are alike. You **Contrast** when you look at how things are different. These pages talk about two kinds of chemical reactions. Compare the two kinds of chemical reactions by writing one way they are alike.

____

Contrast the two kinds of chemical reaction by writing one way they can be different.

____

____

2. Into what elements does water break down?

____

Making New Substances

Reactions of elements and compounds create all the kinds of matter in the world. Sometimes a compound breaks apart to make new substances. It breaks into its elements. In a chemical change, water breaks into the elements hydrogen and oxygen.

In another kind of reaction, two elements *combine*. They come together to make a compound. Hydrogen combines with oxygen to make water. This is the opposite of the first reaction.

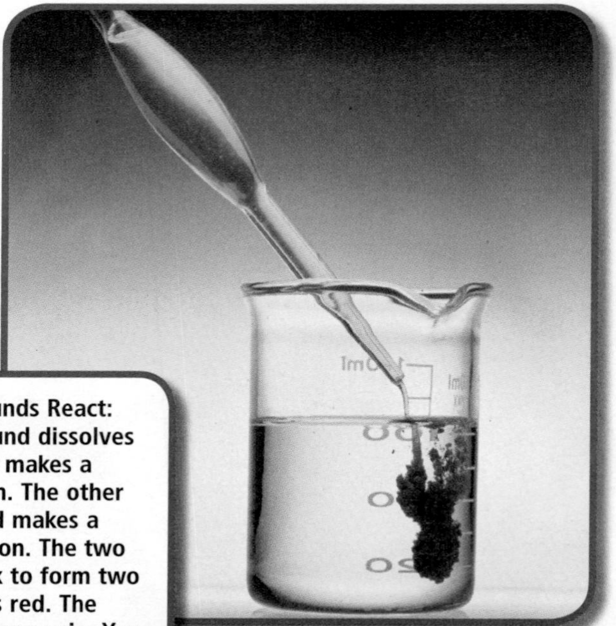

Two Compounds React: One compound dissolves in water and makes a clear solution. The other dissolves and makes a yellow solution. The two solutions mix to form two solids. One is red. The other dissolves again. You don't see it.

▲ **Baking bread is a chemical change. A new substance is formed.**

Elements and compounds combine in other ways to form new compounds. Many carbon compounds have long chains of molecules. They look like paper clips linked together. The chains are found in plastics, proteins, and DNA.

✓Concept Check

1. Compare ways that compounds combine.

____

2. Baking bread is a chemical change. How can you tell?

____

____

3. In what two ways are plastics, proteins, and your DNA alike?

1. ______________________________

2. ______________________________

✓Concept Check

1. You **Compare** when you look at how things are alike. You **Contrast** when you look at how things are different. Compare the atoms of reactants and products in a reaction by naming one way they are alike.

____

Contrast reactants and products in a reaction by writing one way they are different.

____

2. Underline the sentences that tell about the *Law of Conservation of Mass.*

3. Show the mass of the product.

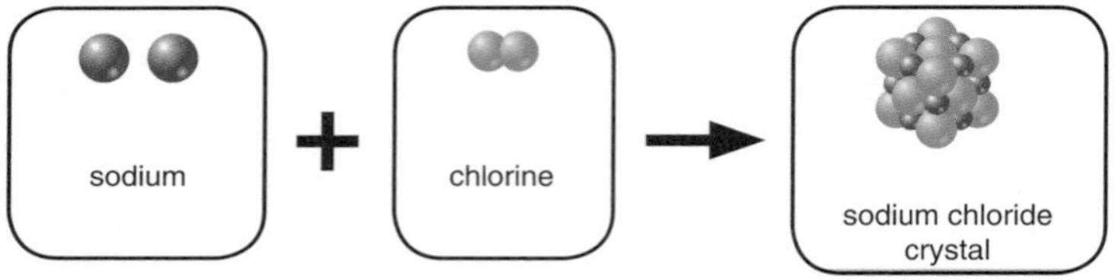

sodium
chlorine
sodium chloride crystal

2 grams + 3 grams → ______

Cooking may change food. But it does not create or destroy it.

Conservation of Mass

During a chemical change, new substances form. These substances are not new matter. The same atoms that were in the reactants are in the products. They are just combined in different ways.

You can measure the mass of the sodium and chlorine gases used in a reaction. After they have changed to salt, you can measure the salt's mass. The masses will be the same. No new matter has been made. No matter was destroyed. This is called the *Law of Conservation of Mass.*

If you burned a marshmallow in a sealed box, you would not lose matter. You could measure mass before and after your test. The marshmallow is smaller but gases were created.

In any chemical reaction, no new matter is formed. Matter is *conserved*. The amount of matter remains the same.

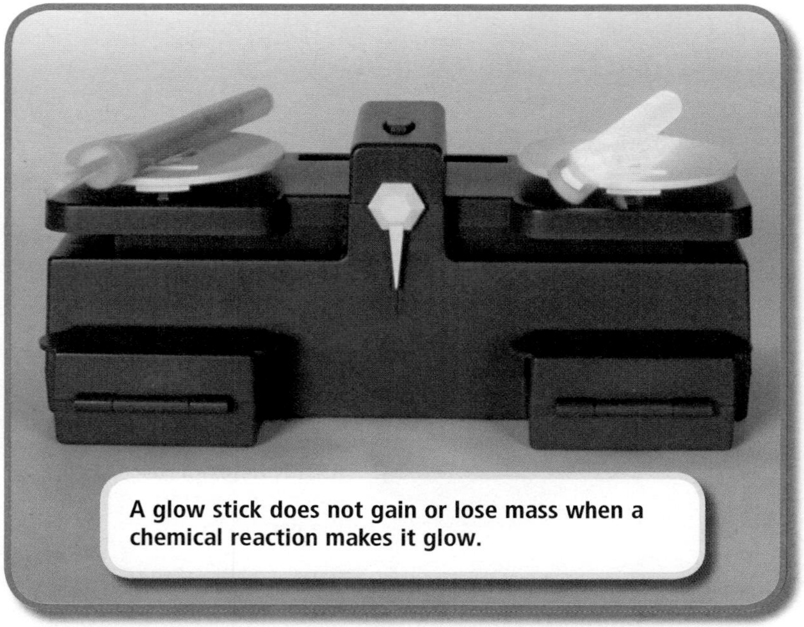

A glow stick does not gain or lose mass when a chemical reaction makes it glow.

✓Concept Check

1. Compare the masses of products of a chemical reaction to the masses of the reactants.

____

2. If you burn a marshmallow in a sealed box, will you lose matter? How would you prove your answer?

____

3. The table below describes what happens during a chemical reaction. Put a check mark in each of the correct columns.

	Changes	**Stays the Same**
Mass		
Atom arrangement		
Amount of matter		
Substance		

✓Concept Check

1. You **Compare** when you look at how things are alike. You **Contrast** when you look at how things are different. Compare acids and bases by writing one way they are alike.

____

Contrast acids and bases by writing one way they are different. Look on pages 6 and 7 for help.

____

2. Underline all of the acids mentioned on this page.

3. Circle any bases mentioned on this page.

Salts

Litmus paper shows when something is an acid or a base. Orange juice is an acid. So is the vinegar used in salad dressing.

Household products like ammonia and soap are bases. So are detergent, shampoo, and drain cleaner. Strong bases are dangerous. So are strong acids. They can burn skin.

When you mix vinegar and baking soda, you get bubbles. A chemical property of acids and bases is that they react with one another.

A car battery has very strong acid in it. This helps make electricity to start the car.

Salt is the product formed when an acid and a base react. There are many salts. Table salt is just one kind of salt.

Most salts are made when strong acids react with strong bases. Salts also form when some metals react with strong acids.

BAKING SODA

Baking soda is a base.

Lesson Review

Complete these Compare and Contrast statements.

1. In ____________ changes, the properties of the materials change.
2. In one kind of chemical reaction, a compound breaks apart. In another, ____________ combine.
3. ____________ changes don't create bubbles, but chemical changes do.
4. ____________ are present at the start of a chemical reaction, and products are present at the end.

Circle the letter in front of the best choice.

1. Which unit has all of the properties of a single element?

   A molecule

   B compound

   C element

   D atom

2. Which is larger than an atom?

   A molecule

   B neutron

   C electron

   D proton

3. Which is a chemical property?

   A color

   B temperature

   C solubility

   D mass

4. Which is TRUE of physical changes?

   A They cannot be reversed.

   B They form new substances.

   C They can be reversed.

   D They involve mixing acids and bases.

5. Which is NOT a property of metals?

   A conduct heat well

   B electricity passes through easily

   C are malleable

   D cannot be rolled or hammered

6. You are putting wiring in your house. Which material would be BEST to use?

   A copper

   B gas

   C carbon

   D hydrogen sulfide

7. What can be done to make a metal stronger?

   A Burn it.

   B Freeze it.

   C Combine it with a nonmetal.

   D Combine it with water.

8. What is air made of?

A gases

B liquids

C solids

D alloys

9. Why is water unlike any other compound?

A It is found in liquid form.

B It is found in solid form.

C It is found in gas form.

D It is found in all three forms.

10. What element is NOT found in the human body?

A Hydrogen

B Bromine

C Oxygen

D Carbon

11. What tool can be used to separate iron from a mixture?

A

B

C

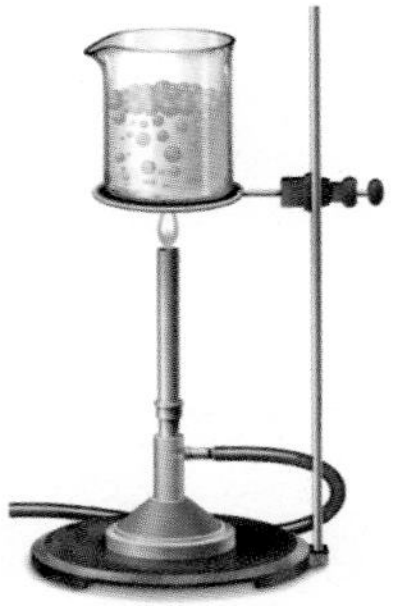

D

12. Which is a possible pH for an acid?

A 3

B 7

C 12

D 16

13. What is happening in this picture? How can you tell?

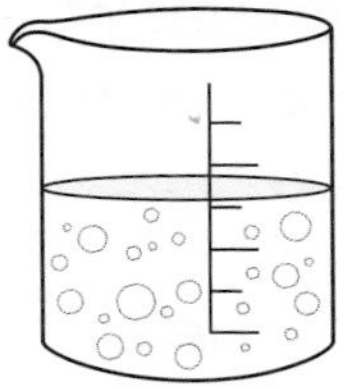

____

____

14. Look back to the question you wrote in your Study Journal. Do you have an answer for your question? Tell what you learned that helps you understand elements and compounds.

____

____

____

Structures of Living Things

In this unit, you will learn how organisms transport materials; how the circulatory, respiratory, and digestive systems work; and how plants and animals rid themselves of wastes. You will also learn how materials are transported through vascular plants and how cells get the energy they need.

Thinking Ahead

How do the different parts of your body get what they need to work?

What are the names of three systems in your body?

Label the parts of the digestive system you know.

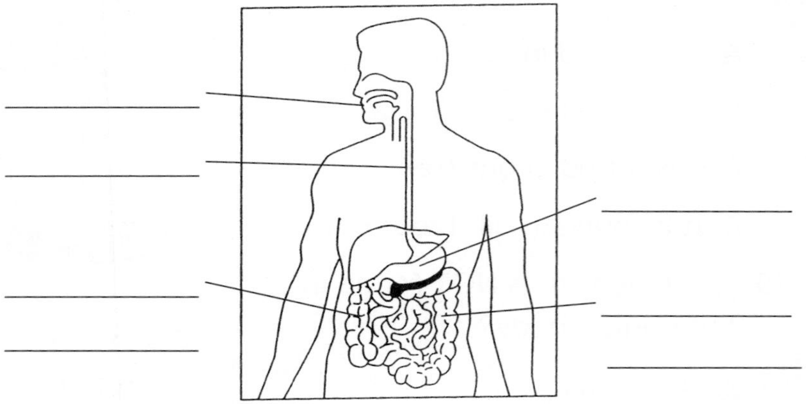

Write a question you have about the structures of living things.

Recording What You Learn

On this page, record what you learn as you read the unit.

Lesson 1

How do organisms transport materials?

___

Lesson 2

Draw a picture of the heart in the box. Show the path blood takes as it enters and leaves the heart. Make the path blue when the blood is carrying carbon dioxide. Make it red when the blood is oxygen-rich.

Lesson 3

Write the names of the parts of the digestive system in order, starting with the mouth and ending with the large intestine.

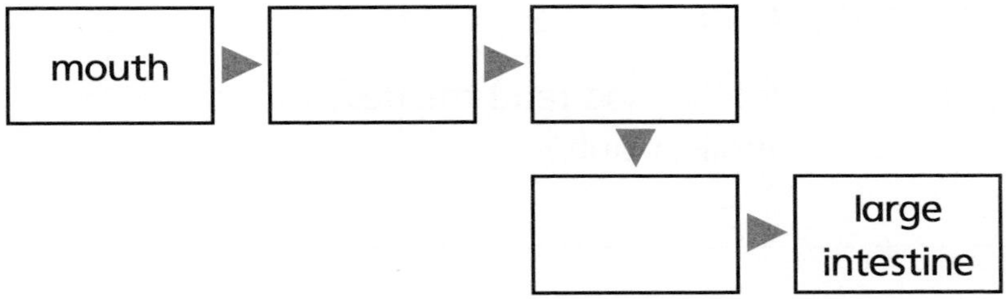
mouth
large intestine

Lesson 4

What wastes do plants produce?

___

Lessons 5 and 6

What are tree trunks made of? How do they help the tree?

___

___

California Standards in This Lesson

2.a *Students know* many multicellular organisms have specialized structures to support the transport of materials.

Vocabulary Activity

Body Systems

The terms name parts of your body that work together to make your body function. See how the parts combine to form bigger parts.

1. Look at the pictures and read the descriptions. Which is the smallest unit?

____

2. Fill in the chart. Look at the pictures for help.

Tissues are made of	
Organs are made of	
Organ systems are made of	

Lesson 1

How Do Organisms Transport Materials?

VOCABULARY
cell
tissue
organ
organ system

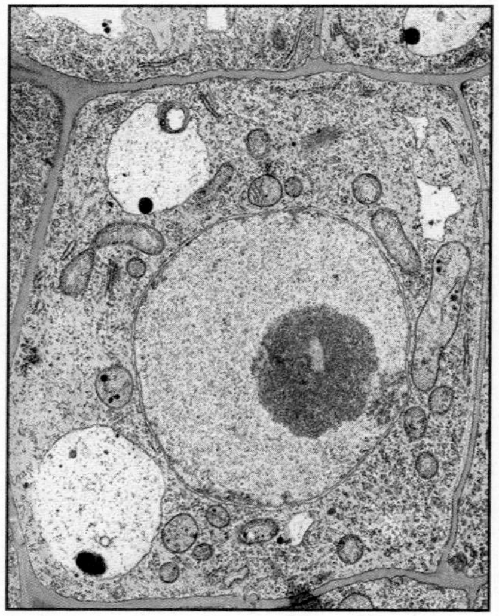

A **cell** is the basic unit of all living things. Most cells can be seen only with a microscope.

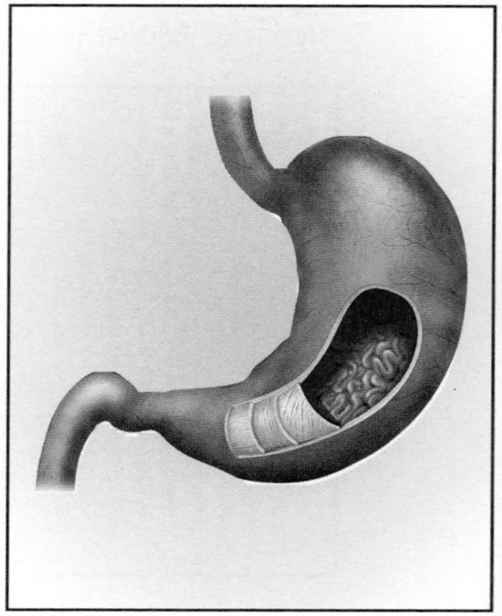

Stomach **tissue** is a group of stomach cells that work together to perform the stomach's function.

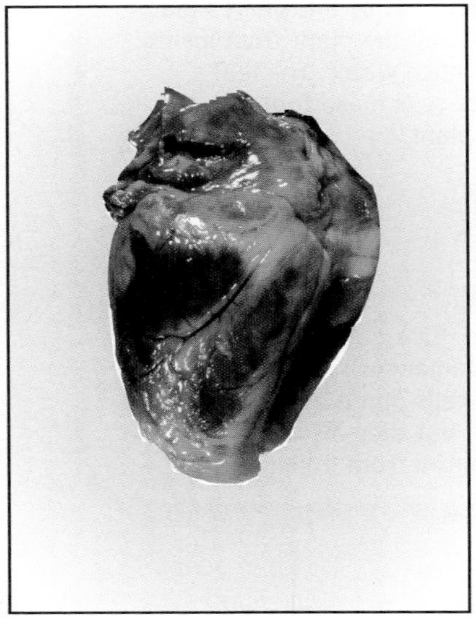

The heart is an **organ** made up of heart tissue.

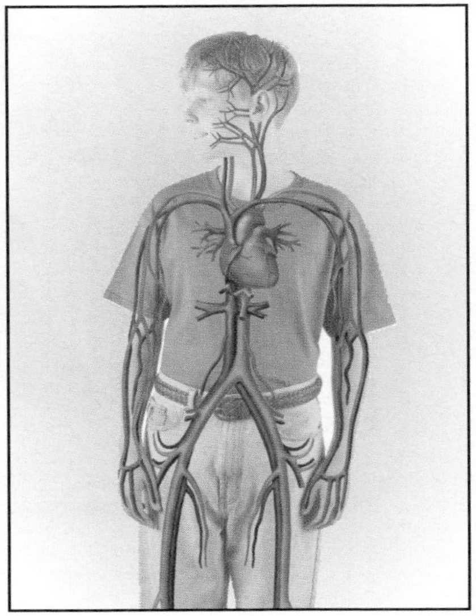

The parts of an **organ system** work together to do a job for the body.

Hands-On Activity
Make a Model

Make a model of either an animal cell or a plant cell. In a bowl, make a gelatin mix. Add pieces of fruit or candy to model different organelles. Set it in the refrigerator to settle.

1. What part of the cell are you modeling with gelatin?

____

2. Did you model an animal cell or a plant cell?

____

3. Which organelles did you include in your cell?

____

____

____

____

✓Concept Check

1. The **Main Idea** on these two pages is A cell is the basic unit of structure and function in living things. **Details** tell more about the main idea. Find details about the role of cells in living things. Underline two of them.

2. What is one detail about cells?

____

3. How can you see a cell?

____

4. Name one job of a cell found in a plant leaf.

____

____

Cells As Building Blocks

A **cell** is the basic unit of structure and function in living things. Every living thing is made up of cells.

Some living things have only one cell. Most living things have many cells. Most cells can be seen only with a microscope.

Plants and animals have different types of cells. Each type of cell has its own job. The cells work together to keep an organism alive and healthy. Each cell contains structures called *organelles*. Organelles have important jobs that keep the cell alive.

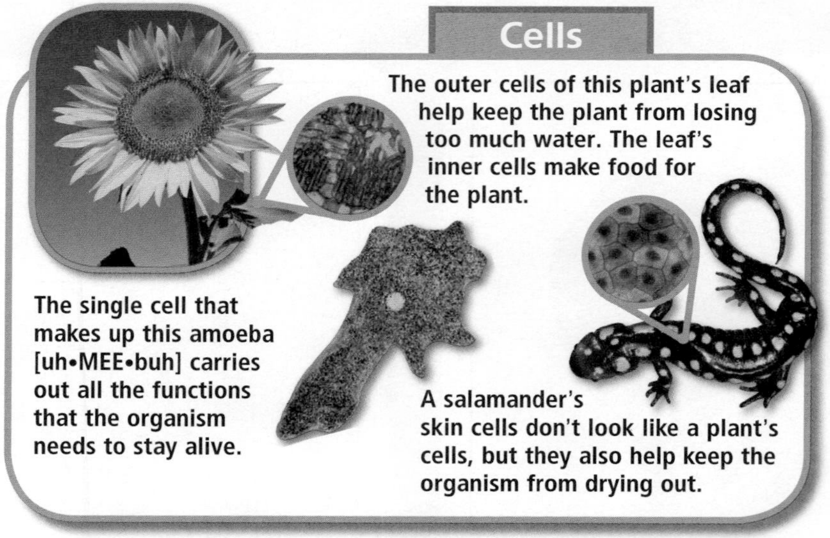
Cells
The outer cells of this plant's leaf help keep the plant from losing too much water. The leaf's inner cells make food for the plant.
The single cell that makes up this amoeba [uh•MEE•buh] carries out all the functions that the organism needs to stay alive.
A salamander's skin cells don't look like a plant's cells, but they also help keep the organism from drying out.

Cell Structures and Functions

All plant and animal cells have certain organelles in common. A cell membrane is an organelle that keeps a cell together. A nucleus is another organelle. It directs all of a cell's activities.

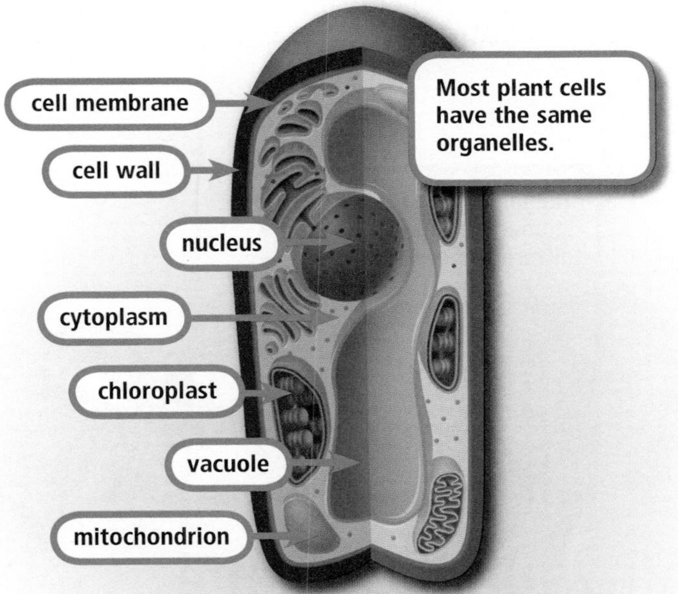
cell membrane
cell wall
nucleus
cytoplasm
chloroplast
vacuole
mitochondrion
Most plant cells have the same organelles.

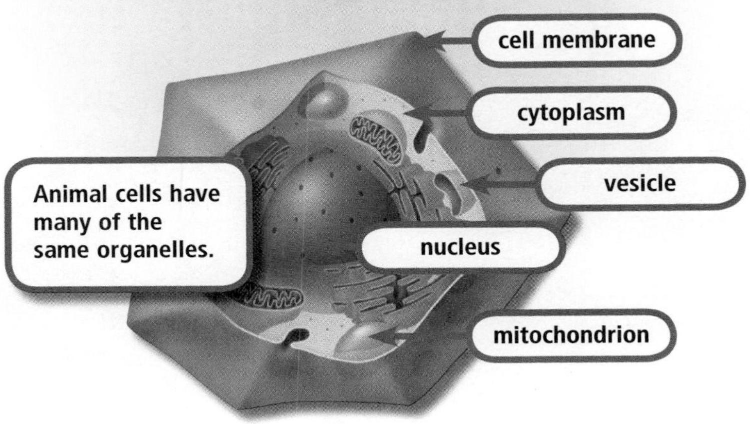
cell membrane
cytoplasm
vesicle
nucleus
mitochondrion
Animal cells have many of the same organelles.

✓Concept Check

1. What organelle directs all the functions of a cell?

____

2. Look closely at the two cells. Circle the names of any organelles that one cell has and the other does not.

3. Tell what each organelle below does.

Organelle	Job
Nucleus	
Cell membrane	

✓Concept Check

1. The **Main Idea** on these two pages is Cells, tissues, organs, and systems work together. **Details** tell more about the main idea. Find details about how these things work together. Underline two of them.

2. Label the organs in the picture on page 6.

3. Does the circulatory system remove cell wastes by itself? Explain.

____

____

4. What is carried in plant and animal circulatory systems?

____

Cells, Tissues, Organs, Systems

Many-celled organisms like you have trillions of cells. **Tissue** is formed by cells that work together for a certain job. Tissues work together in organs. An **organ** is made up of different kinds of tissues working together. Every organ has a certain job in the body.

Different organs also work together in certain ways for the body. Organs that work together form an **organ system**.

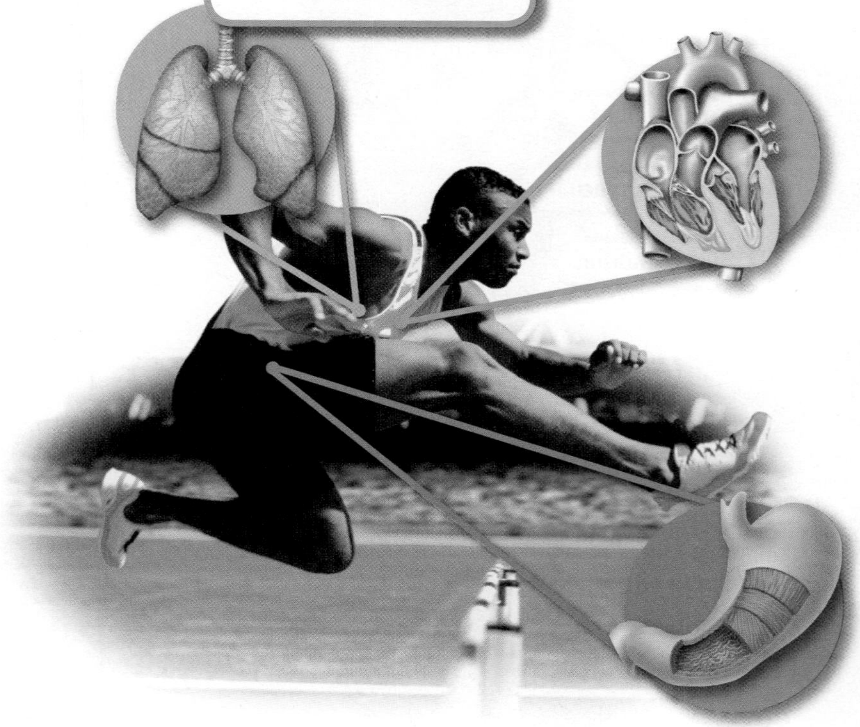

The lungs, heart, and stomach are part of different organ systems.

Transport in Multicellular Organisms

All cells in plants and animals need certain things to live. They need oxygen, water, nutrients, and food. Most plants and animals have structures to bring these things to each cell.

The structures of the circulatory system bring oxygen and nutrients to cells. The circulatory system also works with other organ systems. They work together to remove wastes made by cells.

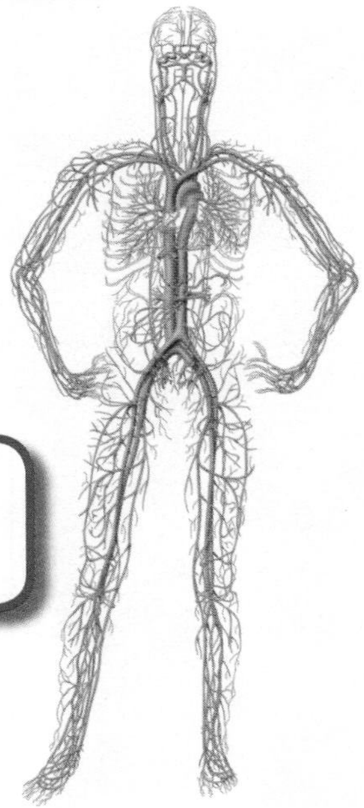

The circulatory system helps bring food to cells and remove waste.

Lesson Review

Complete this Main Idea statement.

1. All living things are made up of ___________.

Complete these Detail statements.

2. Most cells can only be seen with a ___________.
3. Cells have a ___________ that directs all of the cell's activities.
4. Tissue is formed by ___________ that work together.

California Standards in This Lesson

2.b *Students know* how blood circulates through the heart chambers, lungs, and body, and how carbon dioxide (CO_2) and oxygen (O_2) are exchanged in the lungs and tissues.

Vocabulary Activity

Suffix: *-atory*

The suffix *–atory* means “connected with” or “having to do with.”

1. The root word for *circulatory* is *circum*, which means “around.” What do you think the circulatory system does?

____

____

2. The root of *respiratory* is *respire*, which means “to breathe.” What do you think the respiratory system does?

____

Lesson 2

How Do the Circulatory and Respiratory Systems Work Together?

VOCABULARY
circulatory system
respiratory system

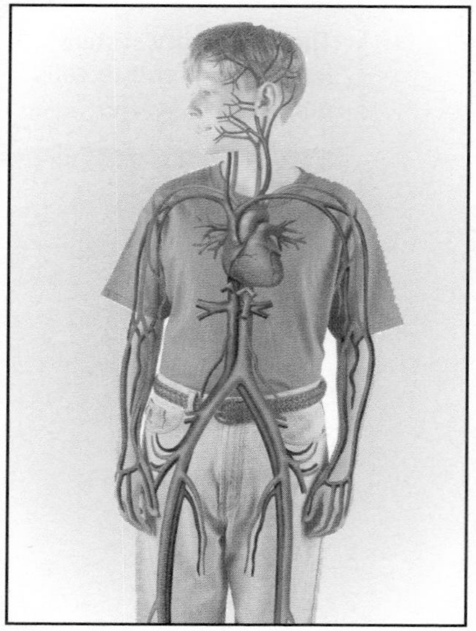

The heart, blood vessels, and blood make up the **circulatory system**.

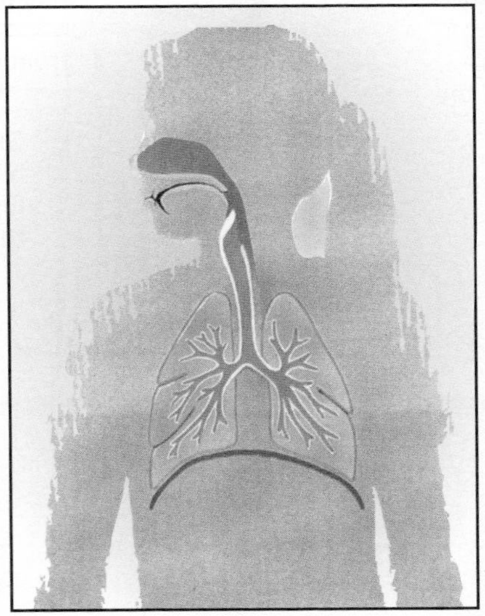

Your **respiratory system** includes your lungs and other organs and tissues. They work together to let you breathe.

Hands-On Activity
The Beat Goes On

1. Find out your heart rate. Place your fingers on a pulse point, such as the side of your neck or wrist, until you can feel your heart beat. Then look at a watch. Count how many times your heart beats in 10 seconds. Multiply that number by 6. This will tell you how many times your heart pumps out oxygen-rich blood each minute.

2. What was your heart rate?

____

3. Now, run in place for one minute. Then find your heart rate again. Did it increase or decrease?

____

4. Why does your heart rate increase with exercise?

____

____

____

✓Concept Check

1. When you put things in a **Sequence**, you put them in order. Underline the first step blood takes when it leaves the heart.

2. Which side of the heart has oxygen-rich blood?

____

3. What happens to blood cells in the lungs?

____

The Heart

The heart is a muscular pump. It carries blood to all parts of the body. Blood travels in a certain path.

Blood from the body returns to the right side of the heart. The heart then pumps it to the lungs. The lungs take carbon dioxide out of the blood. The blood gets oxygen from the lungs. This oxygen-rich blood goes to the left side of the heart. The blood is then pumped to all body cells.

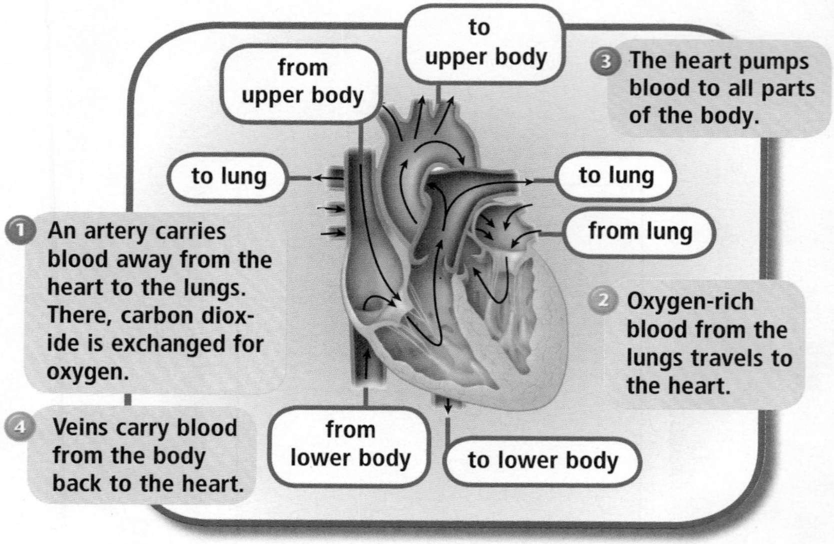

to upper body
from upper body
3 The heart pumps blood to all parts of the body.
to lung
to lung
from lung
1 An artery carries blood away from the heart to the lungs. There, carbon dioxide is exchanged for oxygen.
2 Oxygen-rich blood from the lungs travels to the heart.
4 Veins carry blood from the body back to the heart.
from lower body
to lower body

The Circulatory System

The **circulatory system** is one of the body's most important systems. The circulatory system transports blood loaded with oxygen and nutrients. It brings food to each cell of the body. It also takes away wastes from the cells. The circulatory system is made up of the heart, blood vessels, and blood.

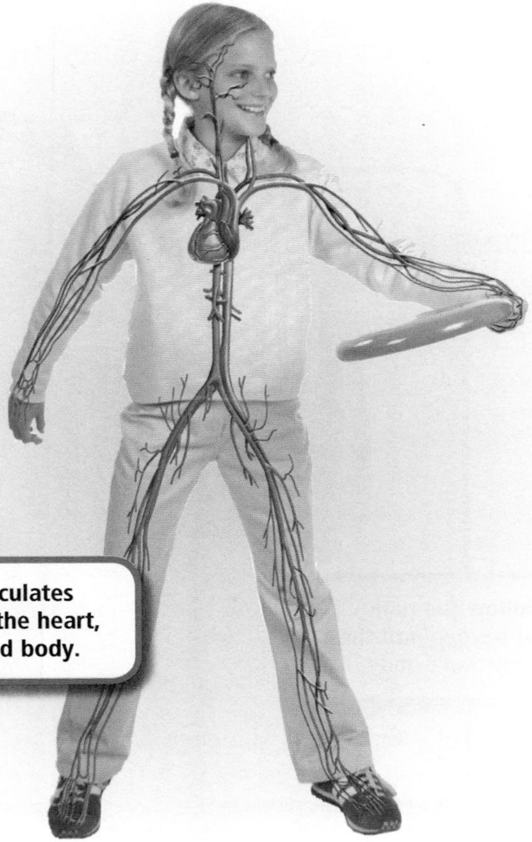

Blood circulates through the heart, lungs, and body.

✓Concept Check

1. What does the circulatory system bring to cells? What does it take away?

2. How do cells get rid of wastes?

3. What system is the heart part of?

4. Circle the things that the circulatory system brings to cells. Underline the things it takes away.

✓Concept Check

1. When you put things in a **Sequence**, you put them in order. Underline the first step blood takes in getting air into the body.
2. What path does air take when you inhale?

____

____

3. How does oxygen get from the air into your blood?

____

4. Look at the picture.

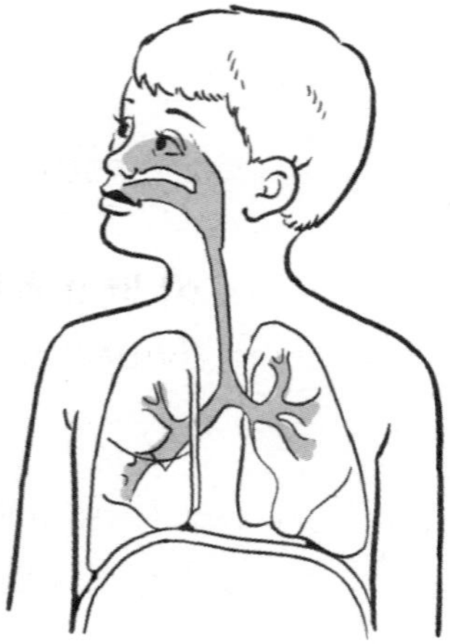

Draw arrows to show the path air takes as it enters and leaves the body. Label the place where the gases are exchanged.

The Respiratory System

Your cells need oxygen. The **respiratory system** makes it possible for the blood to get oxygen for your cells. The respiratory system is a group of organs and tissues. They exchange oxygen and carbon dioxide between your body and your environment.

Air travels from your nose or mouth into the trachea. It travels through smaller and smaller tubes in the lungs. The smallest tubes are where oxygen is exchanged with carbon dioxide.

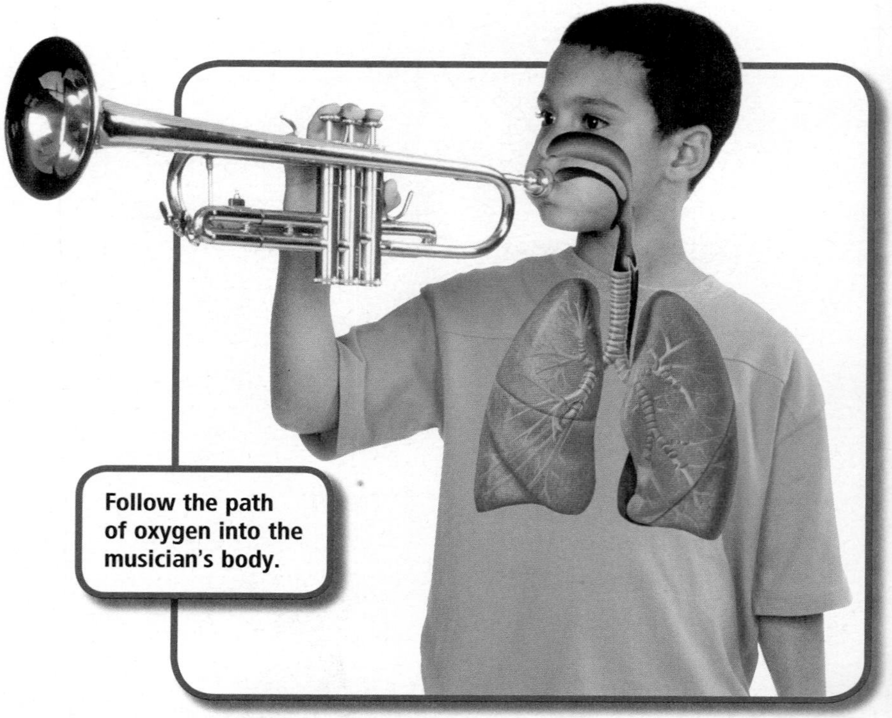

Follow the path of oxygen into the musician's body.

Working Together

Our organ systems are always working together for us. You inhale air and take in oxygen. Oxygen moves from the respiratory system into the blood. Carbon dioxide moves from the blood into the respiratory system. Carbon dioxide leaves the body when you exhale. The exchange of gases takes place in the lungs.

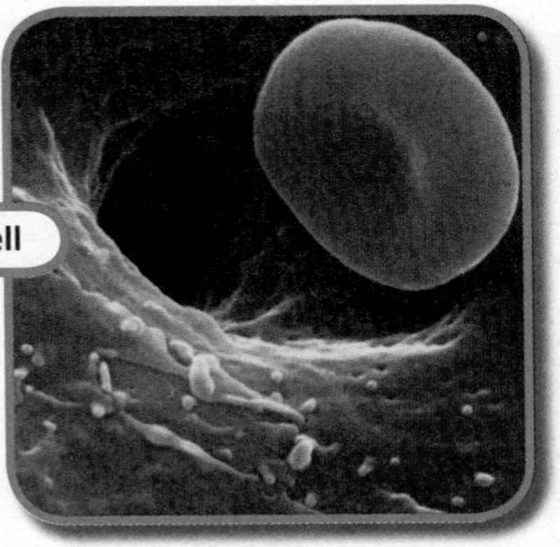

A blood cell

Lesson Review

Complete these Sequence statements.

1. Oxygen enters the respiratory system from the ____________ you breathe in.

2. ____________ from the air you inhale enters the blood.

3. The ____________ ____________ transports blood loaded with oxygen and nutrients to cells.

4. ____________ ____________ leaves the body when you exhale.

California Standards in This Lesson

2.c *Students know* the sequential steps of digestion and the roles of teeth and the mouth, the esophagus, stomach, small intestine, large intestine, and colon in the function of the digestive system.

Vocabulary Activity

Suffix *-ive*

The suffix *–ive* is used to describe something that does or performs an action.

1. What do you think the base word for *digestive* is?

____

2. What job do you think the digestive system performs?

____

____

Lesson 3

VOCABULARY

digestive system

How Do the Organs of the Digestive System Work Together?

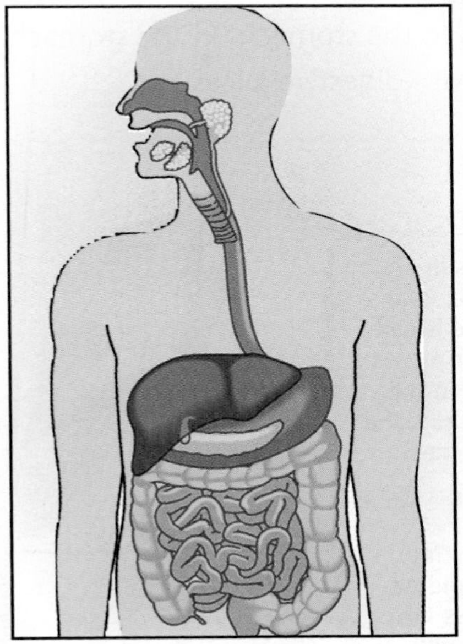

Your **digestive system** breaks down food.

Hands-On Activity
Digestive Action

Use a resealable bag and a cracker to show what the digestion of a cracker may look like.

1. Place the cracker in the bag. Put in a tablespoon of water.
2. Close the bag and crush the cracker with your fingers. Continue crushing it until it is a smooth paste.
3. How does this model digestion?

____

____

____

✓Concept Check

1. When you put things in **Sequence**, you put them in order. Number the steps in the picture below to show the correct sequence of digestion.

2. What path does food follow from the mouth to the stomach?

3. Look at the picture. Circle the organ in which food mixes with digestive juices.

4. What path does food follow from the stomach to the colon?

Food Breaks Down

The organs of the **digestive system** break down food. These organs are the mouth, esophagus, stomach, small intestine, and large intestine.

Your digestive system breaks down food into nutrients. Your cells need nutrients.

Your teeth grind food into smaller pieces. The saliva in your mouth begins the digestion of food. Food then travels down the esophagus to the stomach. In the stomach, strong muscles mix the food with digestive juices.

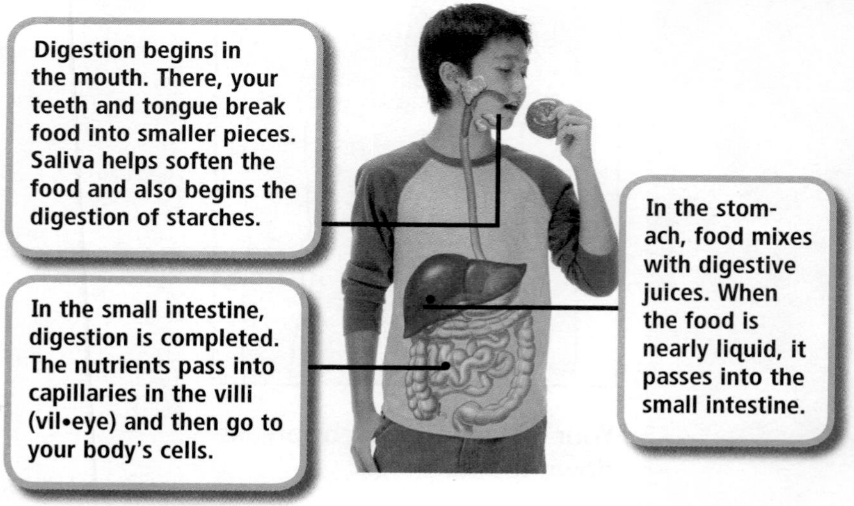
Digestion begins in the mouth. There, your teeth and tongue break food into smaller pieces. Saliva helps soften the food and also begins the digestion of starches.
In the small intestine, digestion is completed. The nutrients pass into capillaries in the villi (vil•eye) and then go to your body's cells.
In the stomach, food mixes with digestive juices. When the food is nearly liquid, it passes into the small intestine.

Water and Nutrients Are Absorbed

Partly digested food flows from the stomach into the small intestine. Chemicals from the liver and pancreas break food down more. Nutrients are absorbed into the blood.

Some food cannot be broken down any further. This food moves into the large intestine. Water from this food is absorbed into the body. What remains is a solid waste called feces. The colon stores it until it is passed from the body.

Lesson Review

Complete these Sequence statements.

1. Your teeth grind food, and then ____________ begins the digestion of food.
2. Food travels down the esophagus to the ____________.
3. Food flows from the stomach into the ____________ ____________.
4. Feces move from the ____________ ____________ to the colon.

California Standards in This Lesson

2.d *Students know* the role of the kidney in removing cellular waste from the blood and converting it to urine, which is stored in the bladder.

Vocabulary Activity

Taking Out the Trash

In this lesson, you learn how your body rids itself of wastes. Some of these words may look new, but you may already know their base words.

1. Use a dictionary to look up the base words given below. Then fill in the chart with your best guess of what each means.

Vocabulary Word	Base	Meaning?
excretory system	excrete	
transpiration	transpire	

2. Look at the picture of the kidney. Why do you think kidney beans are so named?

____

Lesson 4

How Do Plants and Animals Rid Themselves of Wastes?

VOCABULARY
excretory system
kidney
transpiration

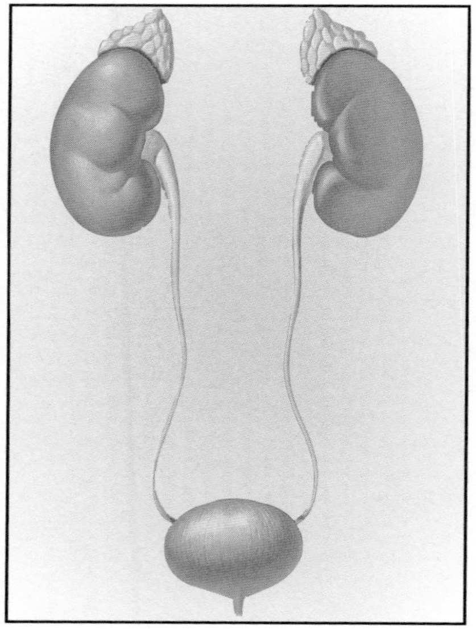

Your **excretory system** removes waste from your body.

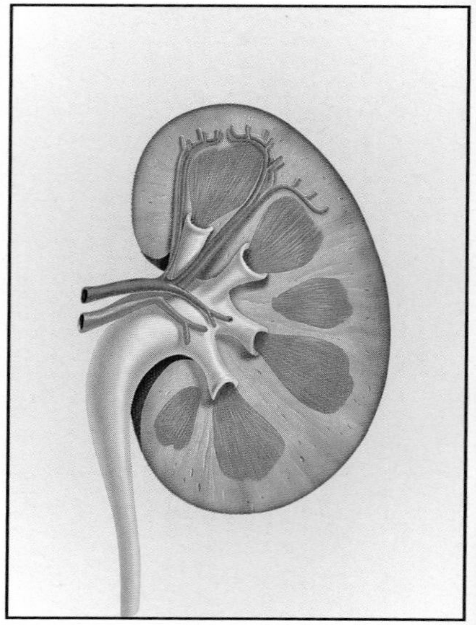

Your **kidneys** are the main organs in your excretory system.

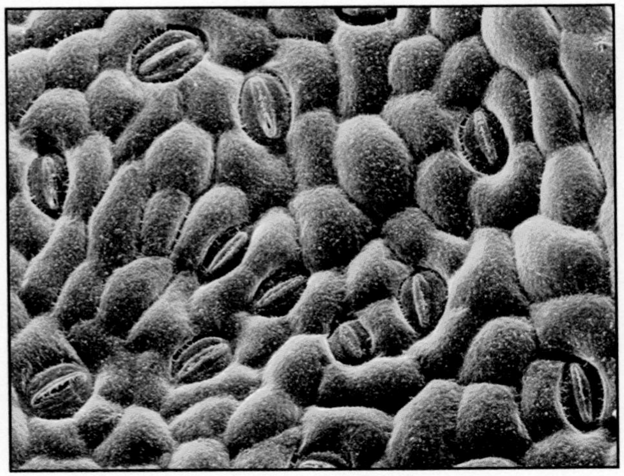

Water moves out of plants through tiny holes in the leaves. This is called **transpiration**.

Hands-On Activity
Water, Water Everywhere

Water the soil in a potted plant. Overwater it a bit. Tie a plastic bag around the plant. Leave it overnight. The next morning, feel the leaves.

1. What did you notice on the leaves?

____

2. Where do you think the water came from?

____

3. Why do you think the plant was getting rid of water?

____

The Excretory System

Your body must remove the things it doesn't need. Carbon dioxide is a waste gas. It is removed from the body as you exhale. Other wastes are removed by the **excretory system**. This system is made up of the kidneys, ureters, bladder, and urethra.

Ammonia is a waste. The blood carries it to the liver. There it is converted to urea. The blood moves it to the bladder as urine. There it is stored until the bladder is full. Then it is removed from the body. Urine flows out of the body through the urethra.

✓Concept Check

1. You **Compare** when you look at how things are alike. You **Contrast** when you look at how things are different. Compare carbon dioxide and ammonia.

   ____

   Contrast carbon dioxide and ammonia.

   ____

   ____

2. Underline the parts of the excretory system.

3. Carbon dioxide is removed differently than other wastes. How is it different?

   ____

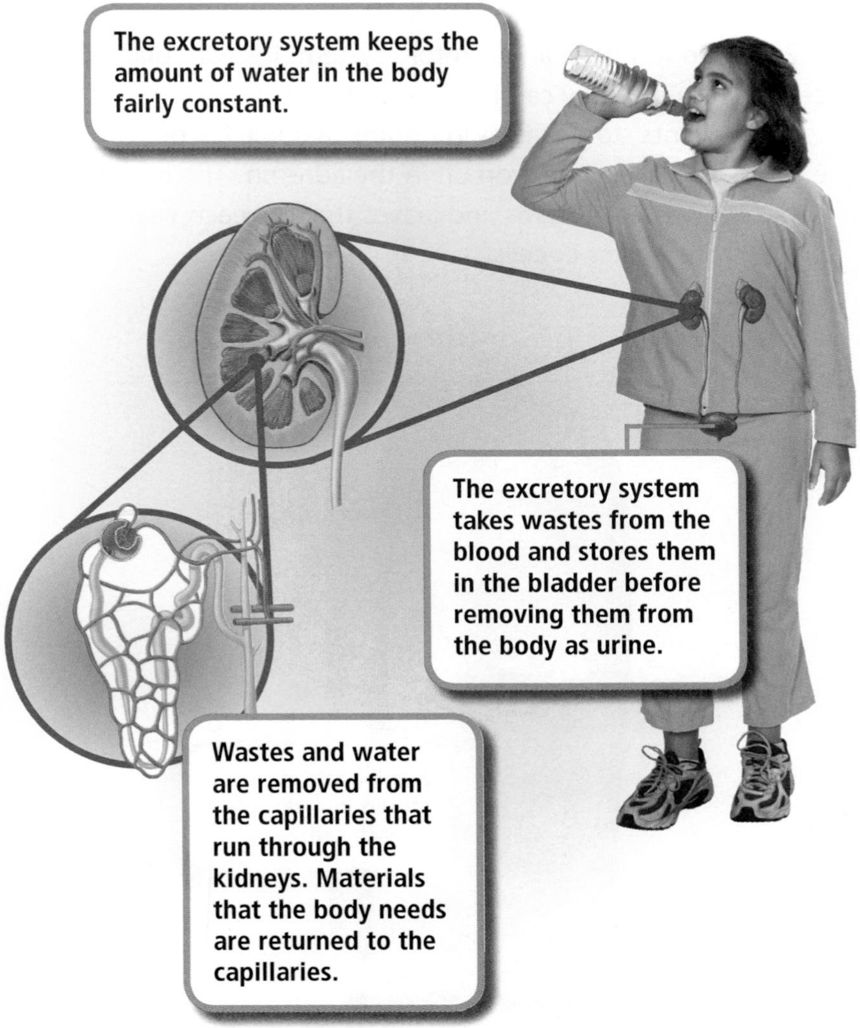
The excretory system keeps the amount of water in the body fairly constant.
The excretory system takes wastes from the blood and stores them in the bladder before removing them from the body as urine.
Wastes and water are removed from the capillaries that run through the kidneys. Materials that the body needs are returned to the capillaries.

✓ Concept Check

1. What organ stores wastes?

____

2. Circle the substance that the excretory system keeps constant. Underline the substances that it removes from the body.

3. Fill in the chart. Tell whether each substance goes to capillaries or comes from capillaries.

Substance	To or From Capillaries?
Materials the body needs	
Water	
Wastes	

✓Concept Check

1. You **Compare** when you look at how things are alike. You **Contrast** when you look at how things are different. Compare the mariposa lily and trees.

____

Contrast how the mariposa lily and trees rid themselves of wastes.

____

____

2. What are plant wastes?

____

3. During transpiration, water moves out of a plant through tiny holes in the underside of leaves. Draw the underside of a leaf during transpiration.

How Plants Rid Themselves of Wastes

Plant cells also produce wastes. Plant waste material includes oxygen and water. Vascular plants have structures that remove wastes.

Plants do not have an excretory system. Instead, plants store cell wastes until they can be removed.

Some plants store wastes in organs that will fall off. An example is leaves that drop off in the autumn. The mariposa lily stores wastes in stems and leaves that die each year. Most plants store wastes in cells.

Plant waste is
stored in a vacuole
in each cell.

Plants do not have lungs to exhale waste gas. They do not have kidneys to filter and remove water. Wastes move out of plants through tiny holes. The holes are on the undersides of leaves. Water moves out of plants in a process called **transpiration**. During transpiration, water moves up through a plant. The water passes through the tiny holes and evaporates.

Lesson Review

Complete these Compare and Contrast statements.

1. Carbon dioxide is removed from the body as you exhale. Other cellular wastes are removed by the ____________ ____________.

2. Like animal cells, plant cells also produce ____________.

3. Unlike animals, plants do not have an ____________ ____________.

4. Animals have kidneys to ____________ waste, but plants do not.

California Standards in This Lesson

2.e *Students know* how sugar, water, and minerals are transported in a vascular plant.

Vocabulary Activity

Word Sounds

The words in this lesson don't look or sound like many other words. However, identifying similar words can help you sound these out.

1. *Xylem* is pronounced *zi-luhm*. *Phloem* is pronounced *flo-em*. Write a word that starts with *xy* that sounds like a *z*. Then write a word that starts with *ph* that sounds like an *f*.

____

____

2. Vascular refers to tubes that carry blood and fluid. What do you think vascular tissue is made of?

____

Lesson 5

VOCABULARY
vascular tissue
xylem
phloem

How Are Materials Transported in Vascular Plants?

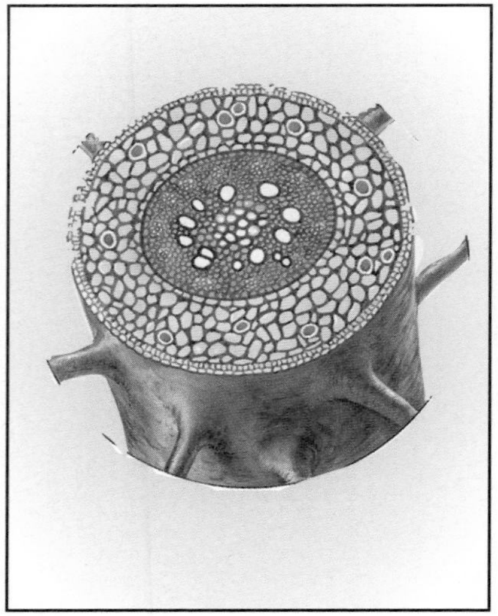

Trees are supported by two kinds of **vascular tissue**. These tubes carry food and water through the plant.

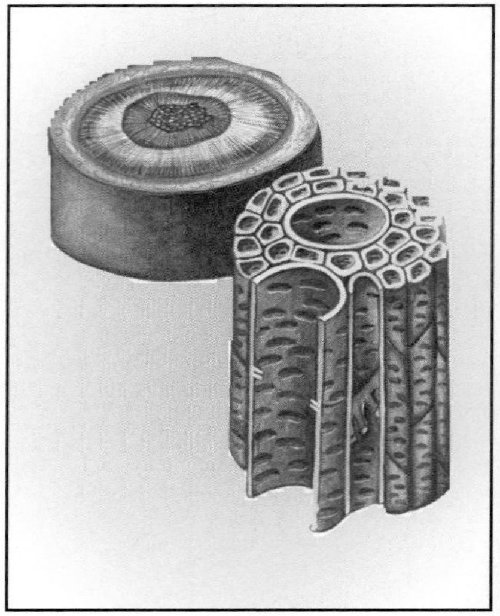

Xylem is a vascular tissue. It carries water and food up from the roots.

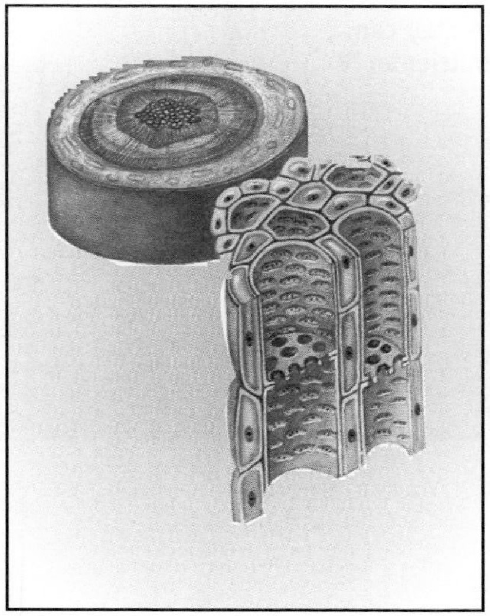

Trees also have **phloem**. Phloem carries food from the leaves to the other parts of a plant.

Hands-On Activity
A Rose of a Different Color

Cut a flower from a plant. Place it in a vase or cup of water. Add several drops of food coloring to the water. Leave the plant in. Record any changes in the plant over the next day.

1. What changes did you notice?

____

____

2. How do you think that happened?

____

____

Concept Check

1. The **Main Idea** on these two pages is There are two types of plants. **Details** tell more about the main idea. Underline two details about types of plants.

2. What transport tissues do vascular plants have?

____

3. Fill in the chart. Check whether each is a property of vascular or nonvascular plants.

Property	Vascular	Nonvascular
No true roots		
Has roots, stems, and leaves		
Absorbs water from surroundings		
Gets water through xylem		
Gets food through phloem		

Nonvascular and Vascular Plants

Mosses are *nonvascular plants*. They do not have true roots, stems, or leaves. They do not have tubes for moving nutrients through them. They absorb water from their surroundings.

Trees are *vascular* plants. They are plants with **vascular tissue**. This tissue supports a plant. It carries water and food. Roots, stems, and leaves all contain vascular tissue.

There are two kinds of vascular tissue. **Xylem** carries water and nutrients up from the roots. It brings them to other parts of a plant. **Phloem** carries food from the leaves to other parts of a plant.

Mosses grow where they can absorb water and nutrients. ▼

New layers of xylem and phloem grow each year.

Xylem cells in the trunk of a tree transport water and nutrients. Phloem cells, just under the bark, transport food. Each year, new layers of xylem and phloem grow. You can tell the age of a tree by counting the rings of xylem.

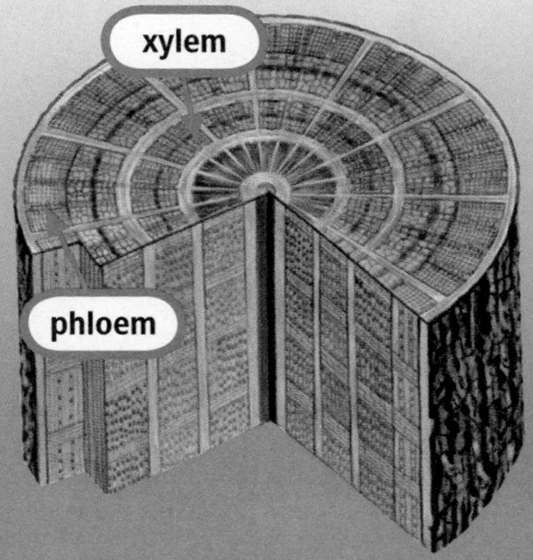
xylem
phloem

✓Concept Check

1. What are two kinds of cells found in tree trunks?

____

2. How often do new layers of xylem and phloem grow?

____

3. Look at the close-up of the tree trunk. Circle the name of the part that transports food. Underline the name of the part that transports water.

4. How can you tell how old a tree is?

____

✓Concept Check

1. The **Main Idea** of these two pages is Roots and stems have vascular tissue. Underline two details about roots and stems.

2. What is the job of roots?

____

3. How do nutrients get from soil to xylem?

____

4. Look at the picture. Label the roots and stem.

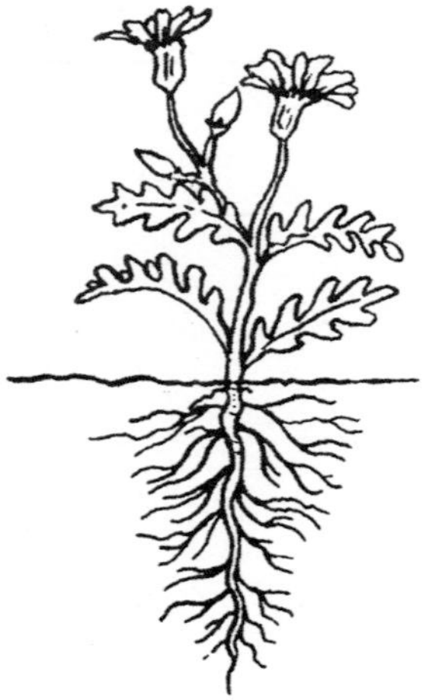

Roots and Stems

Roots and stems have vascular tissue. The tissue absorbs water and nutrients from the soil. How does this happen? Tiny hairs on roots absorb nutrients. Xylem cells take water and nutrients from the root hairs. The xylem cells carry them to the stem.

Roots hold a plant in place in the soil. Like roots, stems have vascular tissue. This helps support plants. The xylem and phloem in stems connect to those same tissues in the roots. Together, they move food for the plant between roots and leaves.

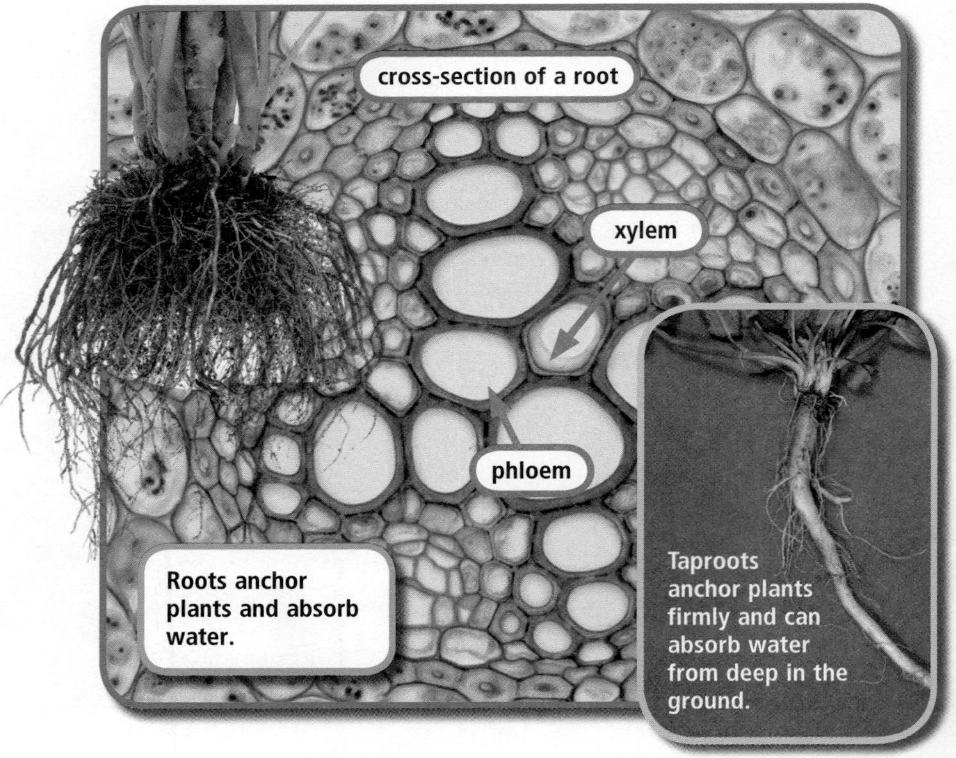
cross-section of a root
xylem
phloem

Roots anchor plants and absorb water.

Taproots anchor plants firmly and can absorb water from deep in the ground.

Vascular tissue grows in rings in tree stems and some other plants. In smaller, softer plants, bundles of xylem and phloem are scattered throughout the stem.

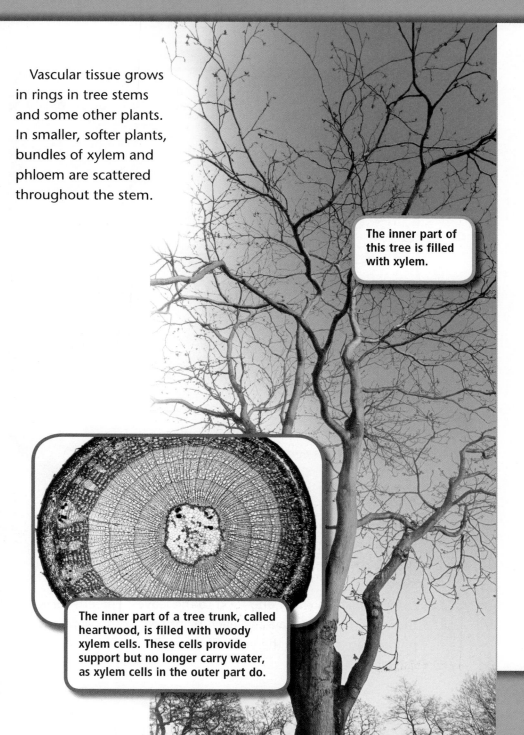

The inner part of this tree is filled with xylem.

The inner part of a tree trunk, called heartwood, is filled with woody xylem cells. These cells provide support but no longer carry water, as xylem cells in the outer part do.

✓Concept Check

1. Name two jobs of xylem cells.

____

2. How does vascular tissue grow in trees?

____

3. Compare and contrast heartwood to outer layers of xylem cells.

____

____

4. Look at the close-up of the tree trunk. How old was the tree it was cut from?

____

✓Concept Check

1. The **Main Idea** on these two pages is Leaves make food for the plant. **Details** tell more about the main idea. Underline two details about how leaves make food.

2. In what part of the leaves is food made?

____

3. Why are leaves green?

____

4. How does sugar get from leaves to every cell in a plant?

____

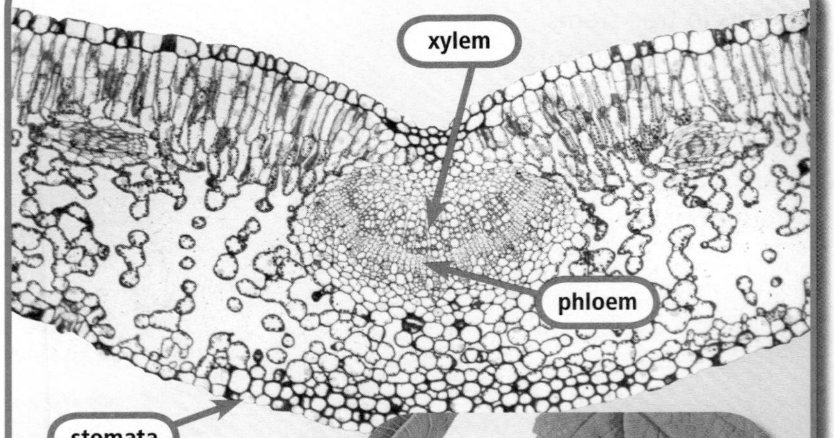
xylem
phloem
stomata

Notice the veins on these leaves. Look at the view through a microscope above.

Leaves

Leaves use sunlight, carbon dioxide, and water to make sugar. Sugar is made inside the leaf cells. Many leaves are green. The green color absorbs the light energy the leaves need.

Xylem carries water and food to the leaves. Phloem carries the sugar made in leaves to each plant cell. Both xylem and phloem are found in the veins of leaves.

eucalyptus tree

Lesson Review

Complete this Main Idea statement.

1. ____________ ____________ carries water and food throughout a plant.

Complete these Detail statements.

2. ____________ carries food and water to leaves.
3. A tree is a ____________ plant. It has tubes to carry nutrients through it.
4. ____________ is a vascular tissue. It carries food from leaves to other parts of a plant.

California Standards in This Lesson

2.f *Students know* how plants use carbon dioxide (CO_2) and energy from sunlight to build molecules of sugar and release oxygen.

2.g *Students know* plants and animal cells break down sugar to obtain energy, a process resulting in carbon dioxide (CO_2) and water (respiration).

Vocabulary Activity

Putting Words Together

You have seen parts of each of these words before.

1. Put these words together and tell their meaning.

cellular + respiration =	
photo (light) + synthesis (to make from)	

2. What is the base word of the vocabulary word fermentation?

____

Lesson 6

How Do Cells Get the Energy They Need?

VOCABULARY
photosynthesis
cellular respiration
fermentation

Photosynthesis helps plants make food and release oxygen.

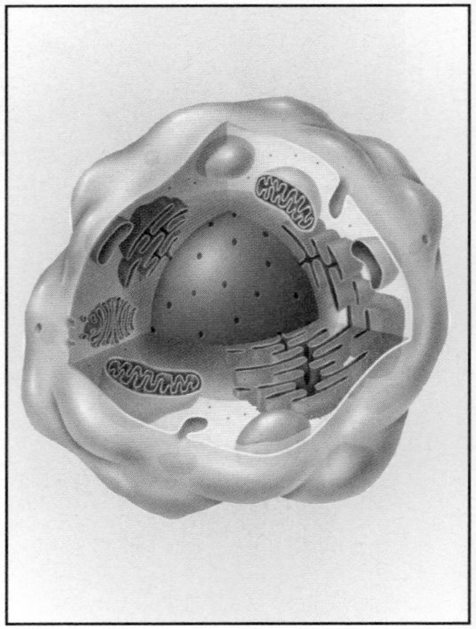

Plants "breathe" through **cellular respiration**.

Fermentation is at work when you make bread. It causes bread dough to rise.

Hands-On Activity
Cotton Roots

Use a cotton ball to show how plants absorb moisture.

1. Slowly dip a cotton ball into a glass of water.
2. Observe as the water is absorbed. The cotton fibers will look like root hairs soaking up water.
3. Why does a plant need roots?

____

____

✓Concept Check

1. When you put things in **Sequence**, you put them in order. Underline the last step in photosynthesis.

2. How do plants get food?

____

3. When does photosynthesis begin? What happens next?

____

____

4. Fill in the chart. Tell what is made during photosynthesis.

Used	Made
Energy	
Carbon dioxide	

Photosynthesis

Photosynthesis produces food for plants. Photosynthesis uses water from the soil. It uses energy from sunlight. It also uses carbon dioxide from the air.

Plant cells have chlorophyll. This allows plants to use light energy to make food. Photosynthesis begins when sunlight is absorbed by plants. This energy breaks down the water in plants. Then carbon dioxide combines with hydrogen from the plants. They combine to form a sugar. This sugar is the plant's food.

Photosynthesis also releases oxygen and water into the air.

Follow the photosynthesis cycle from the sun to plants. ▶

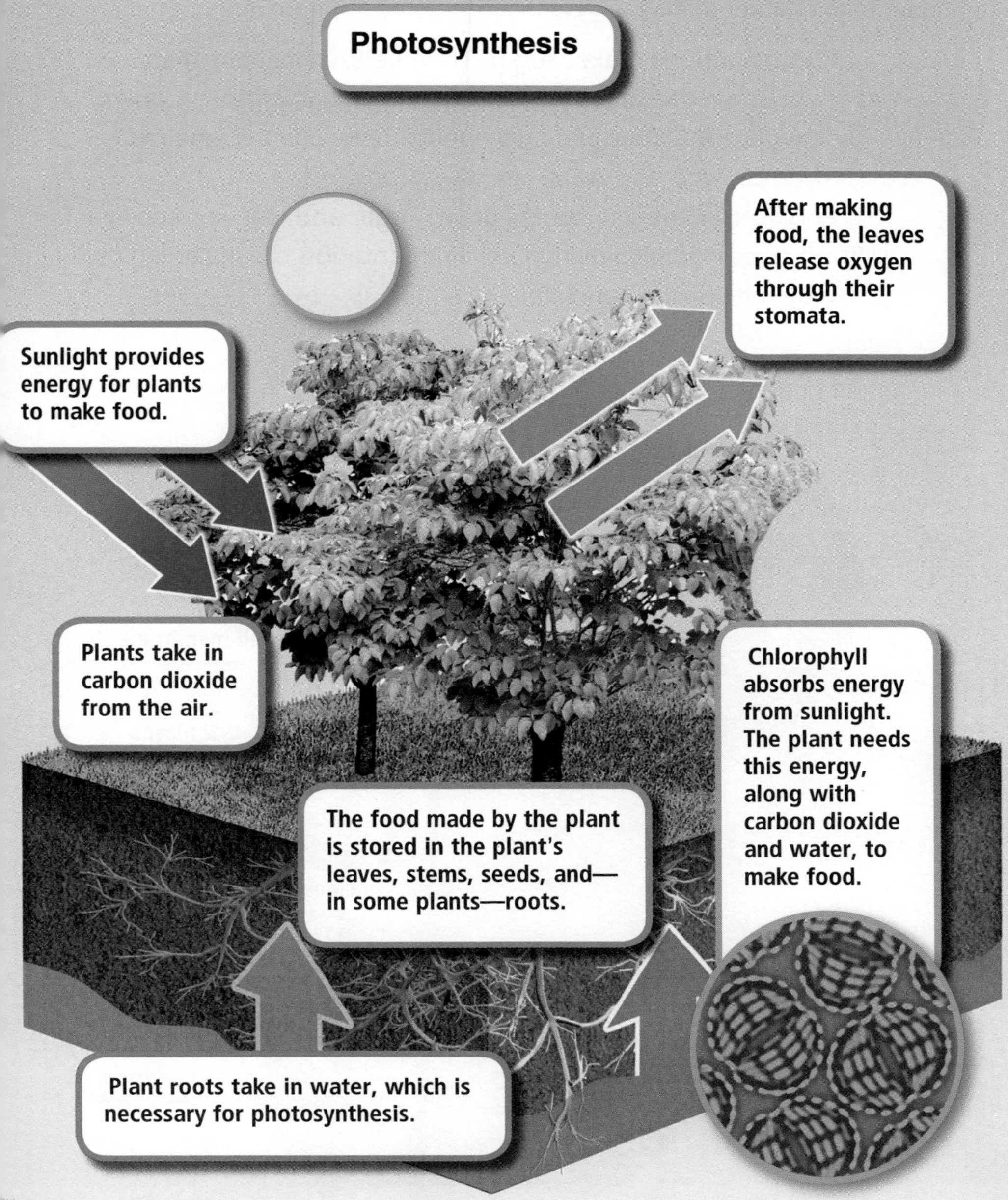
Photosynthesis
Sunlight provides energy for plants to make food.
After making food, the leaves release oxygen through their stomata.
Plants take in carbon dioxide from the air.
Chlorophyll absorbs energy from sunlight. The plant needs this energy, along with carbon dioxide and water, to make food.
The food made by the plant is stored in the plant's leaves, stems, seeds, and—in some plants—roots.
Plant roots take in water, which is necessary for photosynthesis.

✓Concept Check

1. Where do plants get what they need to make food? Match a word on the left with a word on the right.

Energy	**soil**
Carbon Dioxide	**sun**
Water	**air**

2. When is oxygen released during photosynthesis–before or after food is made?

____

3. What does chlorophyll do?

____

✓Concept Check

1. When you put things in **Sequence**, you put them in order. Underline the sentence that tells what happens after photosynthesis makes sugar.

2. What happens to the sugar inside cells?

____

____

3. What causes bread to rise?

____

____

Cellular Respiration

Photosynthesis makes sugar. Then **cellular respiration** breaks down the sugar. Inside cells, the sugar is broken down by oxygen. It is changed into energy. Cells use this energy. Carbon dioxide and water are also produced.

Fermentation also breaks down sugar and releases energy. But it does not require oxygen. Fermentation causes bread to rise by releasing carbon dioxide.

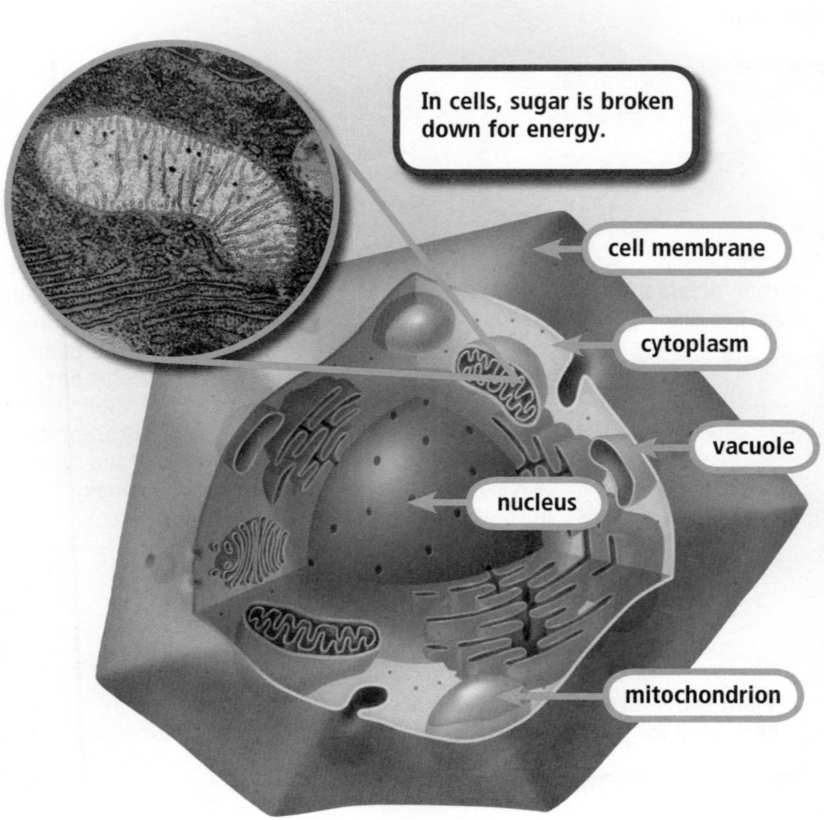
In cells, sugar is broken down for energy.
cell membrane
cytoplasm
vacuole
nucleus
mitochondrion

The Carbon-Oxygen Cycle

Carbon and oxygen are always being recycled through the environment. This cycle depends on two things. These things are photosynthesis and respiration. During photosynthesis, plants take in carbon dioxide. Plants store the carbon from carbon dioxide as food. Then they release oxygen.

During cellular respiration, oxygen helps break down food. Energy is released from food. Also, carbon dioxide is released into the environment.

Lesson Review

Complete these Sequence statements.

1. Photosynthesis begins when ____________ is absorbed by plants.
2. Plants store ____________ as food.
3. Fermentation breaks down ____________ and releases energy.
4. During cellular respiration, sugar is broken down by ____________.

Unit 2 Review

Circle the letter in front of the best choice.

1. Which is the largest?

   A cell

   B tissue

   C organ

   D organ system

2. Which organelle does a plant cell have that animal cells do not have?

   A cell membrane

   B chloroplast

   C mitochondrion

   D nucleus

3. Where does blood get oxygen?

   A heart

   B blood vessels

   C blood

   D lungs

4. Which is NOT part of the respiratory system?

   A nose

   B trachea

   C blood vessels

   D lungs

Use the picture to answer question 5.

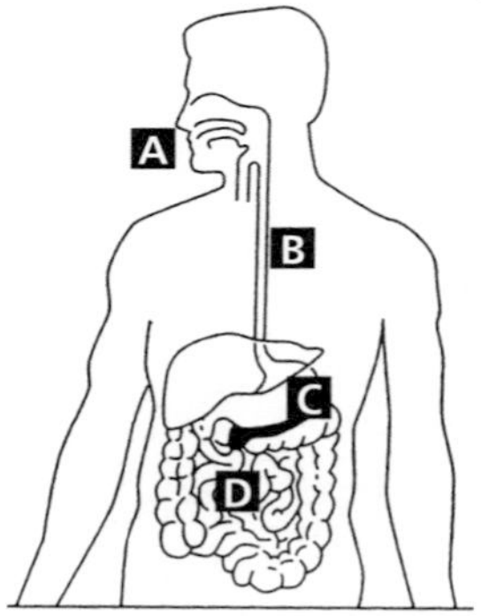
A
B
C
D

5. Where is food mixed with digestive juices?

   A A

   B B

   C C

   D D

6. Which does NOT help break food down?

   A gases

   B teeth

   C juices

   D chemicals

7. Which system is made up of the kidneys, ureter, bladder, and urethra?

   A excretory

   B digestive

   C respiratory

   D circulatory

8. What organ removes waste and water from the blood?

A

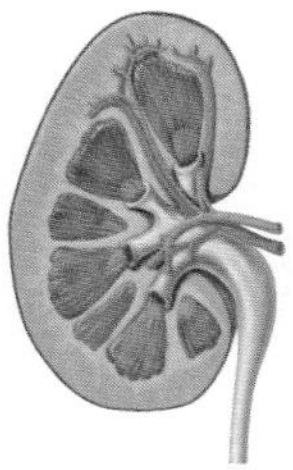

B

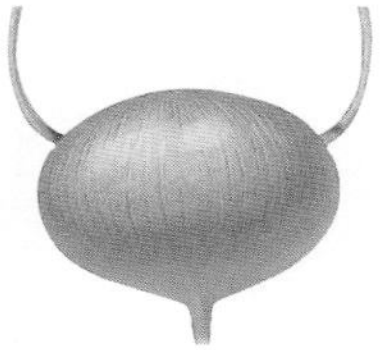

C

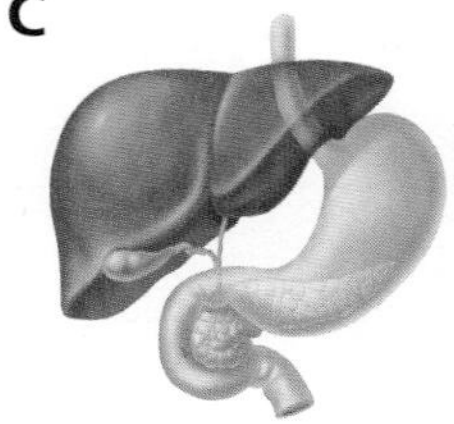

D

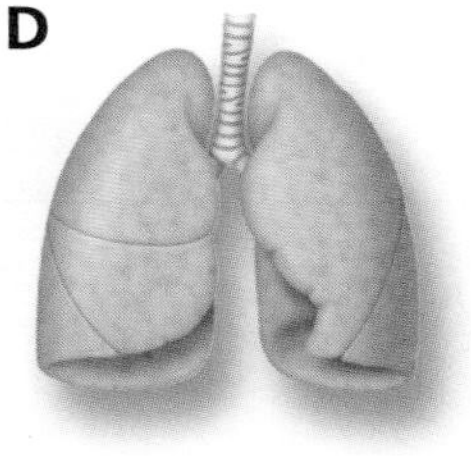

9. What moves water to all of a plant's cells?

A photosynthesis

B transpiration

C xylem

D phloem

10. Which is NOT a step in photosynthesis?

A Sunlight is absorbed by plants.

B Sugar is broken down by oxygen.

C Light energy breaks down the water in plants.

D Carbon dioxide combines with hydrogen from the plants.

11. Suppose there is no oxygen in the air. How can a plant break down sugars?

A It cannot.

B It can perform photosynthesis.

C It can perform transpiration.

D It can ferment the sugar.

12. Where does carbon dioxide in the carbon-oxygen cycle come from?

A photosynthesis

B transpiration

C cellular respiration

D fermentation

13. Look back to the question you wrote on page 48. Do you have an answer for your question? What have you learned about the structures of living things?

____

____

____

____

The Water Cycle

In this unit, you will learn about how water moves from Earth to air and back again, how we get the water we need, and ways to conserve water. What do you know about these topics? What questions do you have?

Thinking Ahead

How does water move from Earth to the air and back again? Draw a picture to show what you think.

What are some ways to conserve water? Write two of your ideas.

Where does the water we use come from? Write what you think.

Write a question you have about the water cycle.

Recording What You Learn

On this page, record what you learn as you read the unit.

Lesson 1

How does water move from Earth to the air and back again? Draw what you learned.

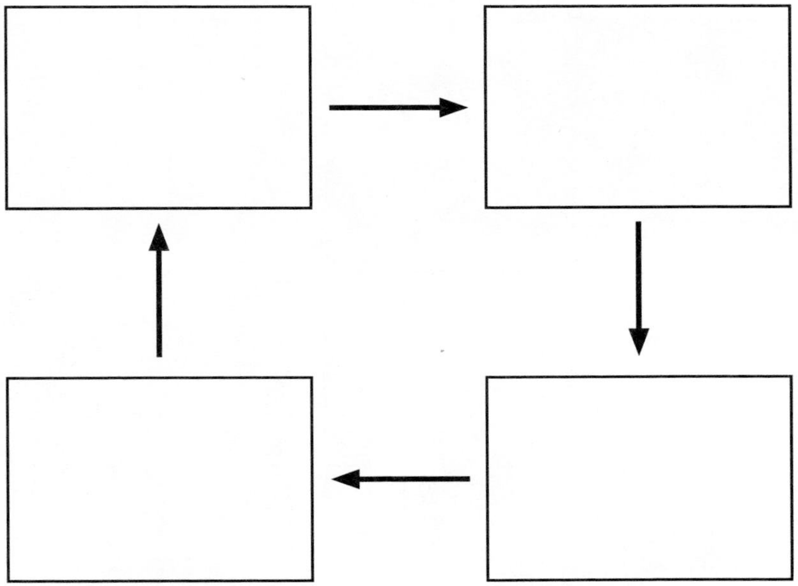

Lesson 2

How do Californians get the water they need? Circle one vocabulary word. Write a fact about it.

watershed **reservoir** **aqueduct** **dam**

___

___

___

Lesson 3

How can people conserve water? Write three ways.

___

___

___

California Standards in This Lesson

3.a *Students know* most of Earth's water is present as salt water in the oceans, which cover most of Earth's surface.

3.b *Students know* when liquid water evaporates, it turns into water vapor in the air and can reappear as liquid when cooled or as a solid if cooled below the freezing point of water.

3.c *Students know* water vapor in the air moves from one place to another and can form fog or clouds, which are tiny droplets of water or ice, and can fall to Earth as rain, hail, sleet, or snow.

Vocabulary Activity

Suffix *tion*

The suffix *-tion* changes the meaning of the base word to which it is added. The suffix *-tion* means "the act of doing something" or "the thing that is being done" or "the state of being a certain way."

1. Circle the two vocabulary words on the page that have the suffix *-tion.*
2. Complete this chart.

Word	Base Word
evaporation	
condensation	

Lesson 1

VOCABULARY

water cycle
water vapor
evaporation
condensation

How Does Water Move from Earth to the Air and Back Again?

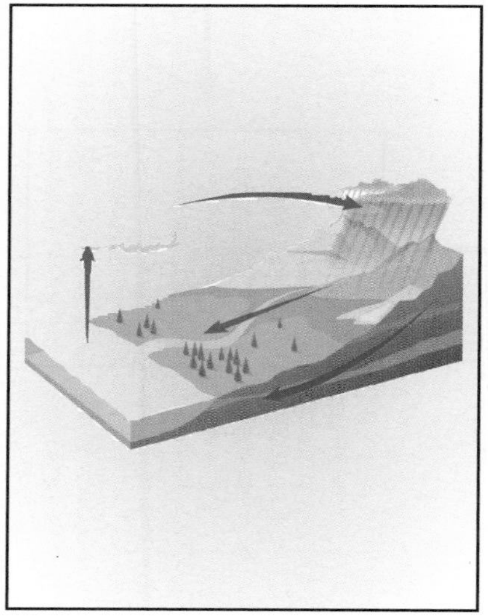

I see a stage of the **water cycle** when rain falls on Earth.

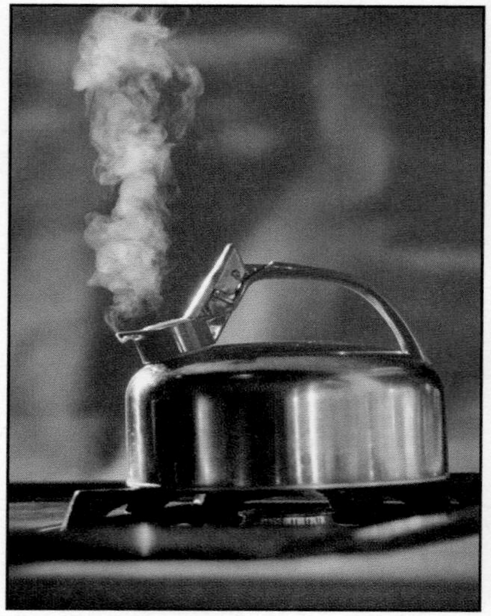

When water boils, it turns into a gas. **Water vapor** is the gas form of water.

When the sun comes out after rain, **evaporation** makes the ground dry again. The water changes into a gas.

You see **condensation** when dew forms on grass. Water vapor in the air changes into liquid water.

Hands-On Activity
Predict, Draw Conclusions

Find a large, clear glass jar with a lid. Fill a small paper cup with salty water. Place the jar lid upside down over the cup, and fit it into the lid. The jar lid does not need to be screwed on.

Predict what will happen.

____

____

____

After 45 minutes, check the jar. What do you see on the walls inside the jar?

____

Taste the water on the jar walls. Is it salty?

____

Draw Conclusions:

____

____

✓Concept Check

1. The **Main Idea** on these two pages is Sometimes Earth is called the water planet. **Details** tell more about the main idea. Find details about why Earth is sometimes called the water planet. Underline two of them.

2. What are the two types of water found on Earth?

____

____

3. Which type is most of Earth's water?

____

4. How are the two types the same and different? Show it on this graphic organizer.

Type of Water	Ways They Are the Same	Ways They Are Different
Salt water		
Fresh water		

The Water Planet

Sometimes Earth is called the water planet. This is because water covers much of Earth's surface. There are two types of water: fresh water and salt water.

Most of Earth's water is in the oceans. Ocean water is salty. Some lake water is salty, too. Most of Earth's water (97 percent) is salt water.

Most of Earth's surface is covered with water. That is why Earth looks blue from space.

The rest of Earth's water—3 percent—is fresh water. People need fresh water for drinking, for washing, and for growing food.

Most fresh water is frozen in ice caps and glaciers. Glaciers are huge sheets of ice. This water is far from most cities and towns. That means it cannot be used by many people.

Fresh water can also be found underground. To get this water, people pump it up to Earth's surface. Other sources of fresh water are rivers and freshwater lakes.

✓ Concept Check

1. Where is most of Earth's water found?

____

2. Find the sentence that tells what people need fresh water for. Circle three ways people use fresh water.
3. How much of Earth's water is salt water?

____

4. How much of Earth's water is fresh water?

____

5. Underline words that tell where fresh water is found.
6. How do people get underground water?

____

✓Concept Check

1. The **Main idea** on these two pages is Water is always moving in the water cycle. **Details** tell more about the main idea. Find details about the water cycle. Underline two of them.

2. What happens when water vapor in the air cools?

____

____

3. What happens to rainwater?

____

____

The Water Cycle

Water is always moving. It moves from Earth's surface to the air. Then the water moves back down to Earth's surface. This never-ending process is called the **water cycle**.

Imagine a puddle of water on the street. Energy, in the form of sunlight, warms the water. This makes the water change state from a liquid to a gas. The gas is called **water vapor**. The water vapor rises into the air.

When water vapor in the air cools, it becomes liquid water again. It forms drops that become heavy enough to fall back to Earth. Most of the water falls into oceans, lakes, and rivers. Some falls onto land. This water soaks into the ground or goes into lakes and rivers.

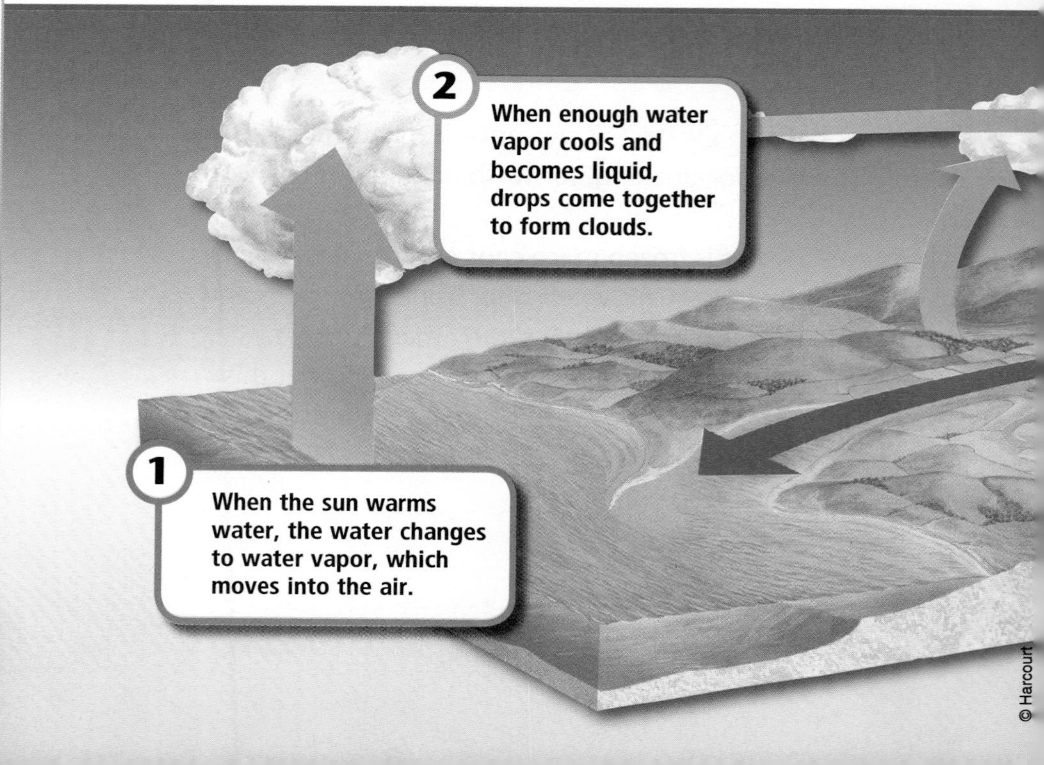

2
When enough water vapor cools and becomes liquid, drops come together to form clouds.
1
When the sun warms water, the water changes to water vapor, which moves into the air.

Some of the fallen water becomes water vapor again as the sun heats it. This is how the water cycle continues.

Most of the water moving through the water cycle comes from the oceans. The energy from the sun heats ocean water and turns it into water vapor. Winds carry the water vapor over land. There it falls as rain, snow, sleet, or hail. Most rainwater comes from the oceans. Why isn't rain salty?

When ocean water turns into water vapor, the salt stays in the ocean. The salt is too heavy to move into the air. Only the fresh water becomes water vapor.

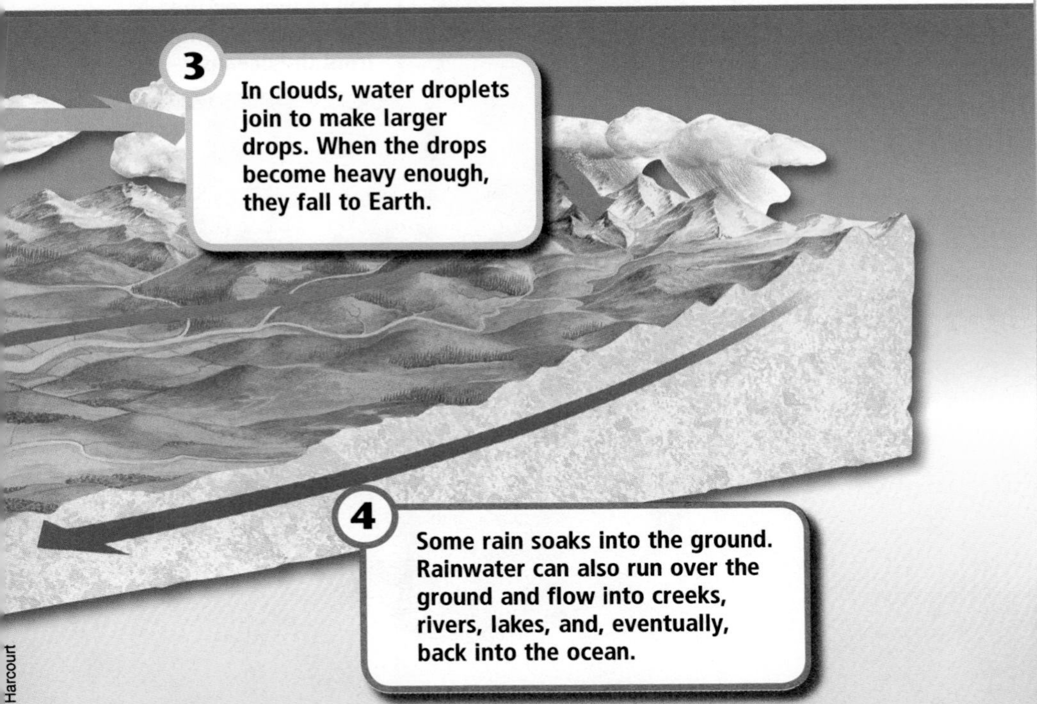

3
In clouds, water droplets join to make larger drops. When the drops become heavy enough, they fall to Earth.
4
Some rain soaks into the ground. Rainwater can also run over the ground and flow into creeks, rivers, lakes, and, eventually, back into the ocean.

✓Concept Check

1. How does water vapor become liquid water again?

____

____

2. Why doesn't salt move into the air when ocean water turns into water vapor?

____

____

Concept Check

1. The **Main Idea** on these two pages is about evaporation. Underline the sentence in the second paragraph that tells the main idea.
2. Several forms of the word *evaporate* appear on this page. For example, *evaporation* uses the base word *evaporate.* Find and circle three different words that are forms of the word *evaporate.*
3. Why is it hard to see evaporation?

____

____

Evaporation

What happens when you lay a damp towel in the sun? The towel dries. Where did the water go?

The water evaporated. **Evaporation** is the process by which a liquid changes into a gas. Water evaporates because heat changes it into water vapor.

It is hard to see evaporation, because water vapor is invisible. Remember the damp towel in the sun? The towel became dry when the water was gone. The water evaporated and became part of the air.

You cannot see the water evaporating from the oceans.

Sweat from your skin can evaporate.

A large amount of water evaporates from Earth's oceans, rivers, and lakes. This happens every day. Water also can evaporate from plants. It evaporates from your skin when you sweat.

Water vapor mixes with other gases in the air. It becomes part of the air.

✓Concept Check

1. What is needed to change liquid water into water vapor?

____

____

____

2. What happens to water vapor?

____

____

____

✓Concept Check

1. Underline the sentence that gives the meaning of the word *condensation.*

2. In the space below, draw and label a diagram of the water cycle.

Condensation

When water vapor becomes part of the air, it moves with the air. Air can carry water vapor very long distances. Water vapor is also carried high into the atmosphere.

As air and water vapor rise, they cool. When the water vapor gets cold enough, it condenses. It changes back into a liquid. **Condensation** is the process by which a gas changes into a liquid.

Water is heated in the teapot. The liquid boils and becomes a gas. The gas is called water vapor. Then the water vapor cools. It condenses into tiny droplets of water as it leaves the pot. This condensed water is called steam.

When water vapor condenses in the air, the liquid mixes with tiny pieces of dust. The water and dust form clouds or fog. Fog is a cloud that forms near the ground.

As more water condenses, the water droplets grow heavier. When the water droplets grow heavy enough, they fall to Earth as rain.

If the air is cold enough, the rainwater freezes. It falls as snow, sleet, or hail.

Condensed water in cold air may fall as snow.

Lesson Review

Complete this Main Idea statement.

1. ____________ is constantly moving from the surface of Earth to the air and then back again.

Complete these Detail statements.

2. Most water on Earth is ____________ water.
3. When water condenses, it changes from a gas to a ____________.
4. ____________ is the changing of a liquid into a gas.

California Standards in This Lesson

3.e *Students know* the origin of the water used by their local communities.

Vocabulary Activity

Californians get the water they need from both natural and human-made sources. Read the vocabulary terms. Then place them in the correct column of the chart.

Natural	Human-made

Lesson 2

How Do Californians Get the Water They Need?

VOCABULARY
- watershed
- dam
- reservoir
- aqueduct
- groundwater

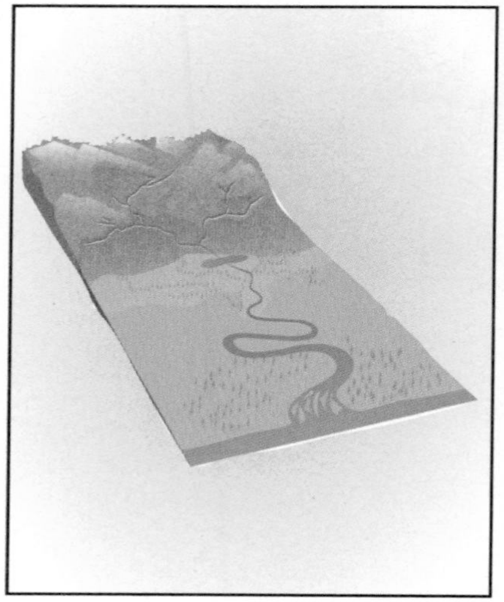

The mountain waters flow down through the creeks and rivers of the **watershed**.

A **dam** helps control a river.

The water we use in our homes was stored in a **reservoir**.

Los Angeles has two large **aqueducts**. They carry water to the city.

When someone pumps the handle on the well, **groundwater** comes out.

Hands-On Activity
Read and Report

Use resources from your local library or the Internet to find out how your community gets its water. Or, report about another city or town near you.

Water Report

Name of city or town:

____

Name of watershed:

____

Is your water stored in a reservoir? Write yes or no. ____

Name of reservoir: ______________________

Does the reservoir have a dam? Write yes or no. ______

Name of dam: ______________________

Do aqueducts transport your water? Write yes or no. __

Name of aqueduct or aqueducts:

____

Draw a picture that shows how water gets to your community.

Concept Check

1. The **Main Idea** on these two pages is Water is useful. **Details** tell more about the main idea. Find details about the uses of water. Underline two of them.

2. Color in the outline below to show how much of the body is water.

3. Why should people drink plenty of water?

____

4. List 5 ways people use fresh water.

 1. ____________________
 2. ____________________
 3. ____________________
 4. ____________________
 5. ____________________

Water Resources

The human body is 60 to 75 percent water. That means you have about 40 liters, or 10 gallons, of water flowing around inside you!

Water is a useful resource. People need water for many different reasons. Fresh water is very important to human health. You need to drink plenty of water to stay healthy.

Fresh water is needed for growing plants for our food. We also use fresh water for bathing, washing clothes, and cooking.

Most crops need fresh water to grow.

Fresh water is important, but salt water is useful, too. Oceans provide many resources.

People who live in places that have very little fresh water may use ocean water. They must first remove the salt and other minerals from the water.

Saltwater animals, such as fish and shrimp, are important to many people's diets. Sea salt is another resource. It is used for cooking and preserving foods.

Oceans are also used for recreation. You can surf, sail, and scuba dive in salt water.

✓Concept Check

1. Where is salt water found?

___

2. What must be removed from salt water so that people can use it?

___

3. Underline three things people eat that come from salt water.

4. Draw a picture showing how the ocean can be used for recreation.

✓Concept Check

1. The **Main Idea** on these two pages is People get fresh water from many places. **Details** tell more about the main idea. Find details about where fresh water comes from. Underline two of them.

2. What keeps lakes and rivers full?

____

3. Draw a circle around the headwaters. Draw a square where the tributary joins the main river.

Local Water Sources

Where do people get fresh water? Many communities in California get water from lakes and rivers. Rain and snow keep lakes and rivers full.

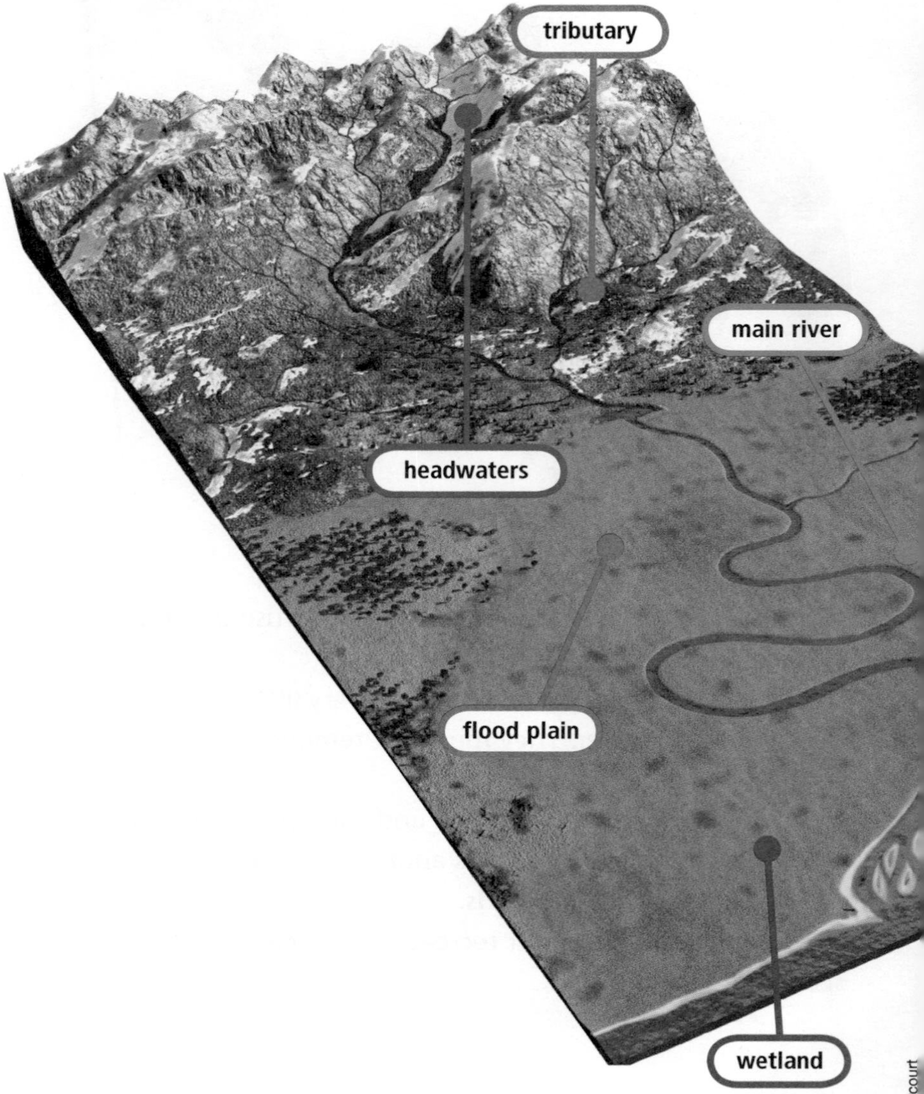
tributary
main river
headwaters
flood plain
wetland

When rain or snow falls, it collects in a **watershed**. A watershed is an area of land with creeks and rivers.

When rain falls or snow melts on a mountain, the water runs into small creeks. The creeks join to form rivers. The creeks and rivers work together to drain the water from an area of land. The area that is drained is called a watershed.

Many communities build dams to store water for future use. A **dam** is a barrier that crosses a river and controls the flow of water. Water collects behind a dam to form a **reservoir**.

California's largest reservoir is Shasta Lake. Its water is cleaned and pumped to homes, businesses, and farms.

If communities need more water, they may bring it from a faraway watershed. They may use an **aqueduct** to transport the water. An aqueduct is a large pipe or channel. Los Angeles uses two aqueducts to bring water from the Owens River valley watershed.

✓ Concept Check

1. Read the statements below. Cross out the statement that does NOT help describe a watershed.
   Creeks join to form rivers.
   Rain falls or snow melts on a mountain.
   A dam is a barrier that crosses a river and controls the flow of water.
   The creeks and rivers drain the water from an area of land.

2. Circle the name of California's largest reservoir.

3. What is the water in Shasta Lake used for?

   ____

4. What are aqueducts used for?

   ____

5. How does Los Angeles get water from the Owens River valley watershed?

   ____

   ____

   ____

Concept Check

1. The **Main Idea** on these two pages is Groundwater is water under Earth's surface. **Details** tell more about the main idea. Find details about groundwater. Underline two of them.
2. What is groundwater?

____

3. What percent of California's population uses groundwater? Circle it.
4. Where does ground water collect?

____

runoff
water table

Groundwater

Not all of California's water comes from rivers and lakes. About 40 percent of California's population uses groundwater. **Groundwater** is water under Earth's surface. It is in the spaces between rocks and soil.

When water soaks into the ground, most of it slowly flows down to an area called the water table. You can think of the water table as a line under the ground.

The area above the water table has pockets of air. The ground below the water table is completely filled with water. It is called an aquifer. It stores groundwater.

Water can be pumped up from an aquifer through a well. A well is a hole that is drilled through the ground to the aquifer. A well can supply water for one home, one neighborhood, or a whole city.

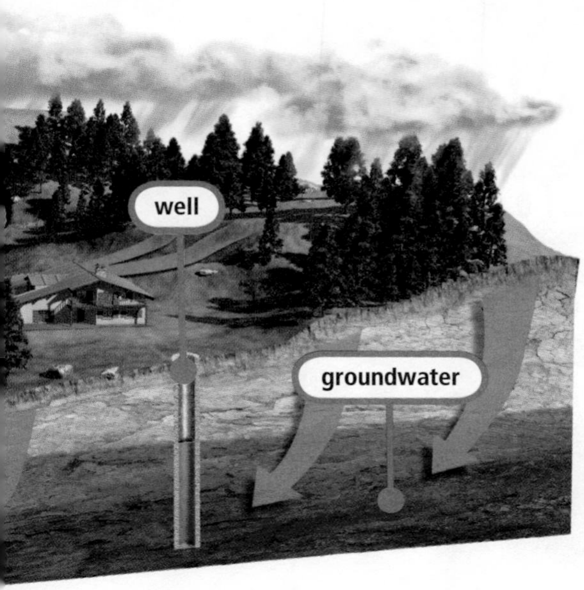
well
groundwater

Lesson Review

Complete this Main Idea statement.

1. Most communities in California get ____________ ____________ from lakes and rivers.

Complete these Detail statements.

2. A community may be able to take fresh water from rivers that drain a ____________.

3. An ____________ can transport water from another watershed.

4. The ground below the water table is called the ____________.

California Standards in This Lesson

3.d *Students know* that the amount of fresh water located in rivers, lakes, underground sources, and glaciers is limited and that its availability can be extended by recycling and decreasing the use of water.

Vocabulary Activity

Descriptions

Draw lines to match the vocabulary terms to their descriptions.

pollution	the practice of recycling and using less water in order to save it
water quality	harmful wastes that enter the water, land, and air
conservation	a good way to conserve water by reusing it
reclamation	a measure of whether water is good to drink

Lesson 3

How Can People Conserve Water?

VOCABULARY
pollution
water quality
conservation
reclamation

Factories sometimes release **pollution** into the water, ground, and air.

DANGER
UNSAFE
WATER

The **water quality** of polluted water is very poor. The water is not safe to drink.

RECLAIMED WATER
USED FOR IRRIGATION
DO NOT DRINK
NO BEBAN EL AGUA
DE IRRIGACIÓN

Reclamation is a good way to conserve water. Farmers use reclaimed water on their crops.

TOWN WATER INTAKE
PLEASE DO NOT WASH
VEHICLES HERE.

When we recycle water and use less water, we practice **conservation**.

Hands-On Activity
Water Conservation Plan

With a family member, write a water conservation plan. Find three ways to conserve water in your home.

1. ______________________________

   ____

2. ______________________________

   ____

3. ______________________________

   ____

Now make a poster showing your water conservation plan. Cut out pictures from magazines or newspapers, or draw your own pictures to illustrate your plan. Hang your poster in your home to remind your family to conserve water.

✓Concept Check

1. A **Cause** is something that makes another thing happen. An **Effect** is the thing that happens. Underline one cause of water pollution. Circle an effect of water pollution in California.

2. What is pollution?

____

____

3. What are three sources of the harmful wastes that cause water pollution?

____

____

____

4. Finish each cause and effect statement.

Waste dumped from __________ __________ __________ pollutes rivers and lakes.

People sometimes use __________ that can pollute groundwater.

If __________ gets into the water that people use, they can get sick.

Water Pollution

People need clean water to stay healthy, but some of Earth's water is polluted. **Pollution** is any change to a resource that makes the resource unhealthful to use.

Water can become polluted when harmful wastes enter the water cycle. Some harmful wastes come from factories or mines. These can be dumped into rivers and lakes. Wastes can also get into groundwater. People who have yards or farms often use chemicals that can pollute groundwater.

Another source of water pollution is sewage. Sewage is human waste. If sewage gets into the water that people use, they can get sick.

Wastes can
spill or leak
The wastes
seep into
undergrou
water.

Scientists measure how safe water is for humans, animals, and plants. This measurement is called **water quality**. Water is safe to use if it has good water quality. Pollution harms water quality. Polluted water is dangerous to use.

Laws are written to help keep good water quality. Factories and cities must clean water after they use it. Also, they can't dump chemicals in rivers.

Groundwater Pollution

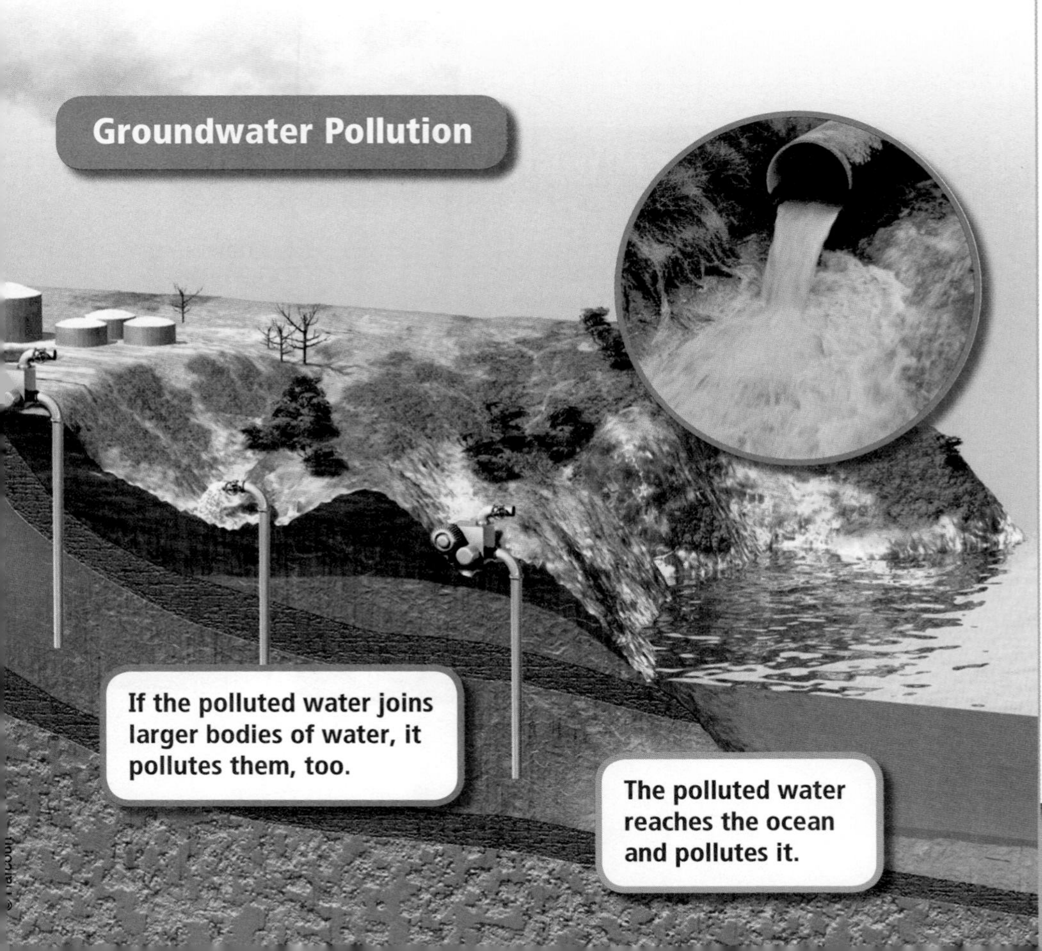

If the polluted water joins larger bodies of water, it pollutes them, too.

The polluted water reaches the ocean and pollutes it.

✓Concept Check

1. Draw lines to match the cause to the effect.

Wastes spill or leak.		Larger bodies of water become polluted, too.
Polluted water joins larger bodies of water.		The ocean becomes polluted.
Polluted water reaches the ocean.		Wastes get into underground water.

2. Can factories dump chemicals into rivers? Why or why not?

____

____

3. Look at the picture on pages 110–111. Draw an arrow pointing in the direction the runoff from rain and snow would go.

✓ Concept Check

1. A **cause** is something that makes another thing happen. An **effect** is the thing that happens. Underline one cause of water conservation.

2. Why does California need to conserve water?

____

____

3. Circle the activities that can use reclaimed water. Cross out the activities that cannot use reclaimed water.

fighting fires

flushing toilets

drinking water

watering plants

Water Conservation

There is limited fresh water in reservoirs and in the ground. Water is a resource that must be conserved, or protected. Water **conservation** is important in California because of droughts. A drought is a long period without rain. Also, the population of California is growing. More people need to share water resources.

We can reduce the amount of water we use. Water can also be recycled. Polluted water can be cleaned at a water treatment plant.

Steps in Waste-Water Treatment

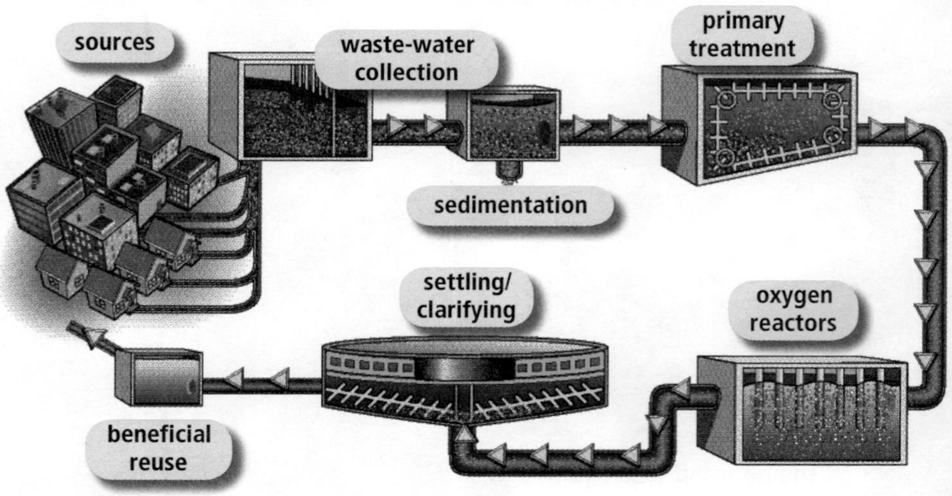
sources
waste-water collection
primary treatment
sedimentation
settling/ clarifying
oxygen reactors
beneficial reuse

Another way to conserve water is not to have a lawn. Lawns need great amounts of water. To save water, homeowners can grow plants that use less water.

Some communities practice water **reclamation**. Reclaimed water is not cleaned as well as drinking water. Reclaimed water can be used for fighting fires, flushing toilets, and watering certain plants. Reclaimed water is not for drinking.

Farmers can conserve water by dripping it onto the ground through small tubes instead of using sprinklers. This uses much less water.

You can conserve water, too. Take shorter showers. Do not run the water while you brush your teeth. Try to think of other ways to conserve our water.

Lesson Review

Complete these Cause and Effect statements.

1. Water becomes ____________ when harmful substances enter the water cycle.
2. Because we have limited amounts of fresh water, it must be ____________ or ____________.
3. To conserve water, ____________ water can be used to water plants, flush toilets, and wash cars.
4. Polluted water can lower the water ____________ of groundwater.

Circle the letter in front of the best choice.

1. A student knows that his family gets its water from a well. What often determines the amount of water that will be available to the family?

   A yearly rainfall and snowfall
   B daily temperature changes
   C distance from the ocean
   D height above sea level

2. What happens to water vapor as it rises higher and higher?

   A The vapor cools and then condenses to form clouds.
   B The vapor heats and then evaporates to form clouds.
   C The vapor heats and then condenses to form clouds.
   D The vapor cools and then evaporates to form clouds.

3. Which two words refer to a barrier across a river and the water stored behind the barrier?

   A aqueduct and dam
   B conservation and reclamation
   C dam and reservoir
   D reclamation and reservoir

4. Which bodies of water contain most of Earth's water?

   A creeks
   B glaciers
   C oceans
   D rivers

5. Which is a period of little rain?

   A drought
   B preservation
   C protection
   D resource

6. What is the name for a pipe or channel used to transport water?

   A aqueduct
   B dam
   C environment
   D runoff

7. How can we conserve water?

   A consume
   B discard
   C pollute
   D recycle

8. Which material that can enter the water cycle is not a source of pollution?

   A fertilizer
   B pesticide
   C rain
   D sewage

9. What is the source of the energy that drives the water cycle?

   A oceans
   B rain
   C sun
   D wind

10. When water evaporates, it turns to a gas. What is the word used to describe this gas?

   A rain
   B sleet
   C snow
   D vapor

11. What word is used to describe the recycling of used water?

   A condensation
   B destruction
   C evaporation
   D reclamation

12. What chemical is left behind when water from the oceans evaporates and becomes fresh water?

   A chlorine
   B rain
   C salt
   D snow

13. What is formed when water from the aquifer comes to the surface naturally?

   A dam
   B spring
   C watershed
   D water table

14. Clouds are made of water droplets. What causes the water to fall to Earth?

   A Droplets move closer together and form a gas.
   B Droplets join to make drops.
   C Droplets become cold and turn to gas.
   D Droplets become warmer and rise.

15. Look back to the question you wrote on page 88. Do you have an answer for your question? Tell what you learned that helps you understand the water cycle.

____

____

____

____

____

____

____

Weather

In this unit, you will learn how the uneven heating of Earth affects weather and how the oceans and the water cycle affect weather. You will also learn about how weather is predicted, and about the causes and effects of severe weather. What do you know about these topics? What questions do you have?

Thinking Ahead

The air around Earth is the atmosphere. Draw a layer of atmosphere around Earth.

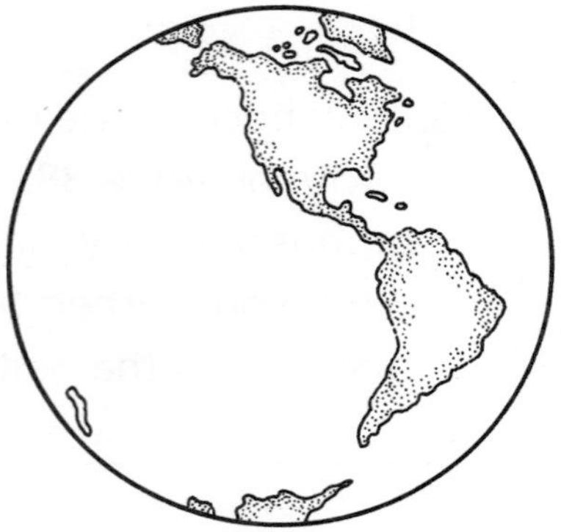

What does the uneven heating of Earth cause? Write what you think.

____

Where does rain come from? Write what you think.

____

What are three tools you can use to measure weather?

____

____

What does a hurricane look like? Draw what you think.

Write a question you have about weather.

____

____

Recording What You Learn

On this page, record what you learn as you read the unit.

Lesson 1

Read about the layers of air in the atmosphere. Order them by how dense the air is in each, with 1 being the densest.

____________ air in the middle of the atmosphere

____________ air closest to Earth

____________ air in the upper atmosphere

Lesson 2

For each number in the diagram, write the event that leads to rain falling as part of the water cycle.

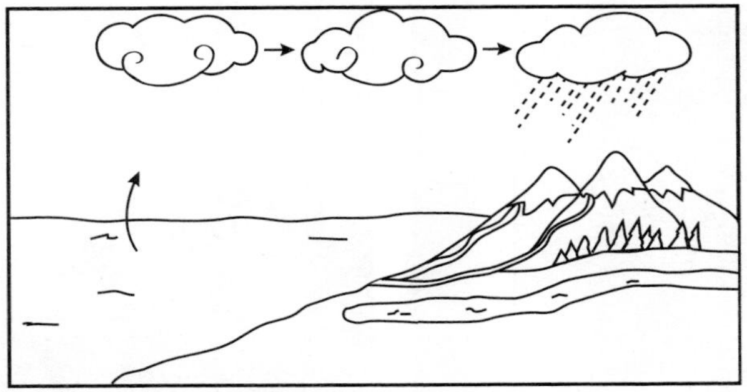

1. ____
2. ____
3. ____

Lesson 3

Tell what each tool measures.

Tool	What It Measures
barometer	
anemometer	
hygrometer	
thermomenter	

Lesson 4

Solve the severe weather riddles.

- I start as a low pressure area over the ocean. I am a ____________.
- I start as a warm air mass above land. I sometimes give off electric charges! I am a ____________.
- I start when winds spin in a column of air. I am a ____________.

California Standards in This Lesson

4.a *Students know* uneven heating of Earth causes air movements (convection currents).

4.e *Students know* that the Earth's atmosphere exerts a pressure that decreases with distance above Earth's surface, and that at any point it exerts this pressure equally in all directions.

Vocabulary Activity

Weather

In this lesson, you'll learn how weather is affected by uneven heating.

1. Some vocabulary terms are made of two words you already know. Break the terms into these words.

air	
	current
	wind
prevailing	

Lesson 1

How Does Uneven Heating of Earth Affect Weather?

VOCABULARY
atmosphere
weather
air pressure
convection current
local wind
prevailing wind

The **atmosphere** is layers of air that cover Earth.

I can watch **weather** as it changes.

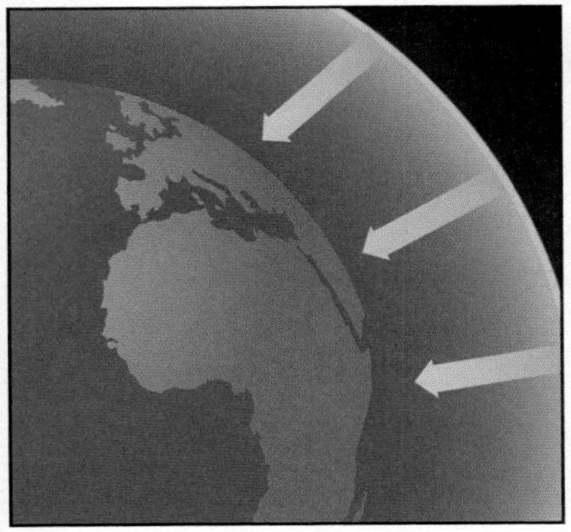

The atmosphere presses down on Earth and causes **air pressure**.

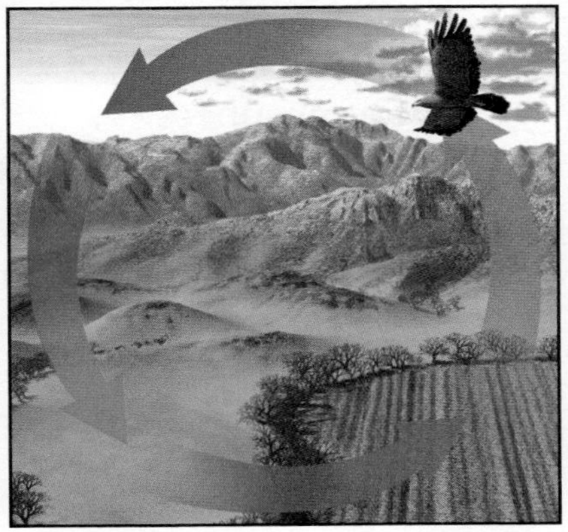

When temperature changes, it causes **convection currents** in the atmosphere, as well as a **local wind** to blow across the land.

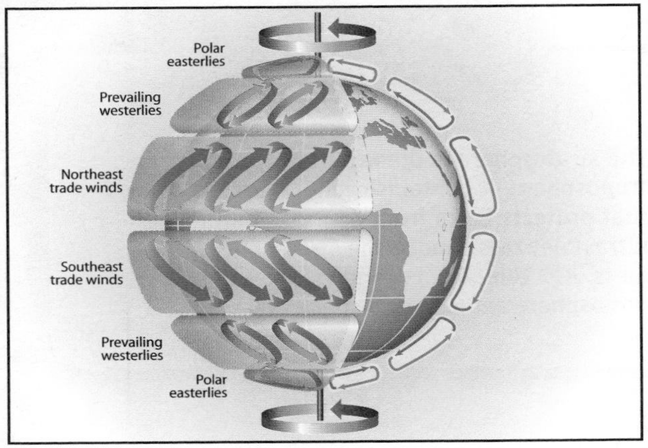
Polar easterlies
Prevailing westerlies
Northeast trade winds
Southeast trade winds
Prevailing westerlies
Polar easterlies

Columbus used the **prevailing winds** to help him sail his ships.

Hands-On Activity
Uneven Heating

1. Fill a large container with very warm water. Fill a small cup with very cold water, and add a few drops of food coloring.
2. Use tongs to gently lower the colder cup of water straight down into the warm water. Make sure the rim of cold water goes below the surface of the warm water.
3. What happens to the colder water?

____

Cool air is denser than warm air, so it sinks. The cool air pushes the warm air up. Draw a picture of this. Use arrows to show the moving air.

✓ Concept Check

1. The **Main Idea** on these two pages is The atmosphere surrounds Earth. **Details** tell more about the main idea. Circle two details about the atmosphere.

2. Look at the picture. Shade the layer of the atmosphere in which weather happens.

3. How far does the atmosphere extend?

____

4. What is the atmosphere?

____

5. Underline the sentence that tells what ozone is.

The Atmosphere

There is a blanket of air that surrounds Earth. The blanket is the **atmosphere**.

The atmosphere is made up of many layers. The layer closest to Earth's surface is the *troposphere* (TROH•puh•sfeer). Most of Earth's weather happens in the troposphere.

Atmosphere Layers

The stratosphere is the layer above the troposphere. It contains ozone. Ozone is a gas that protects Earth from the sun's harmful ultraviolet rays. Above the stratosphere, the air is very thin. The outermost layer of the atmosphere extends into space.

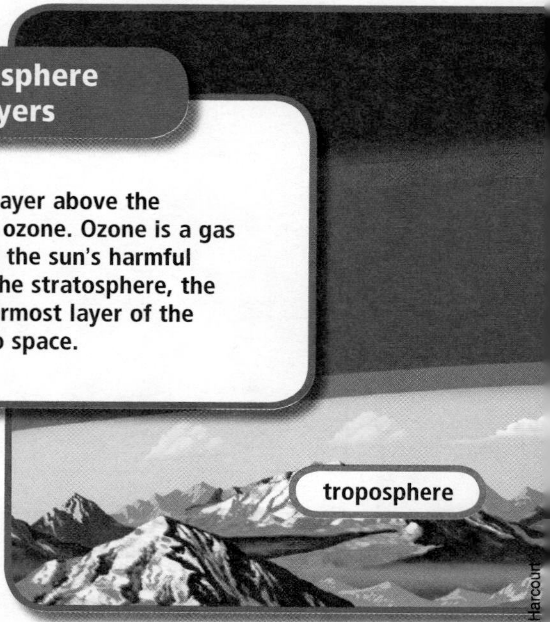
troposphere

Earth's atmosphere experiences many different conditions. We call these changing conditions **weather**. Weather is the condition of the atmosphere at a certain place and time.

You cannot see, taste, or touch air, but you can feel it. When the wind blows, you can feel the air against your face.

Air has mass, and it presses down on Earth. The weight of air in the atmosphere is called **air pressure**.

Air closest to Earth's surface has more weight than air higher up in the atmosphere. Air is denser closer to Earth because of gravity's pull. The dense air causes air pressure to be greater at sea level.

ozone
stratosphere

Concept Check

1. How do you know air is there?

____

2. In the box below, draw the sky. Show where the stratosphere and troposphere are located. Use an arrow and label to show where the densest air is found.

3. Look at page 120. In which layer does most weather occur?

____

4. What do we call the weight of air pressing down on Earth?

____

Concept Check

1. The **Main Idea** of these two pages is The atmosphere has air pressure. **Details** tell more about the main idea. Circle two details about air pressure.
2. Look at the balloons. Use arrows to show the direction in which the balloons expand when you blow air into them.
3. What causes air to press down on Earth?

____

Air Pressure

As you go higher into the atmosphere, there is less air. Less air above you means you feel less air pressure.

You can measure air pressure. Air presses downward because of gravity. Air also pushes in other directions. Blowing air into a balloon shows this idea. As you blow into the balloon, the air pressure increases and the balloon expands. The balloon expands in all directions.

Air pressure expands these balloons in all directions.

Temperature also affects air pressure. Cold air is denser than warm air. Because of this, cold air sinks toward Earth's surface. As it sinks, the cold air forces warmer air to move up. When warm air rises, it begins to cool and become more dense. Then this cooler air sinks back to the surface.

Dense, cold air has high pressure and sinks. Warm, less dense air has low pressure and is pushed up. The areas of high pressure balance the areas of low pressure.

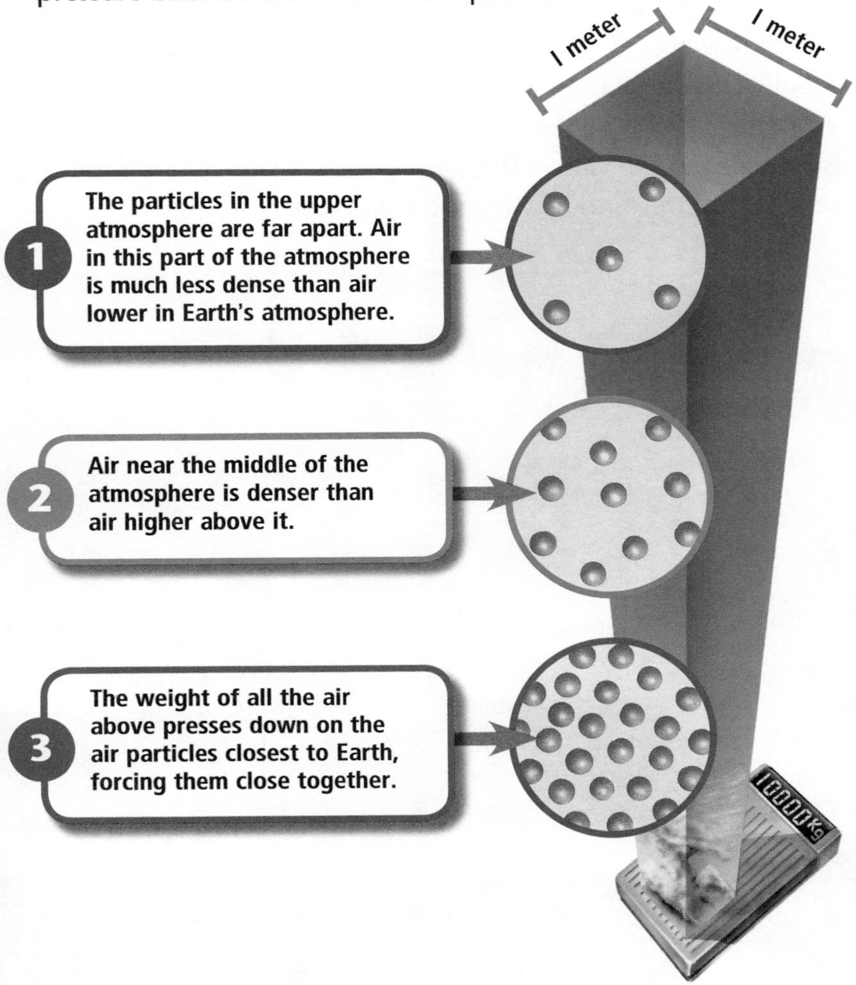

I meter
I meter
1 The particles in the upper atmosphere are far apart. Air in this part of the atmosphere is much less dense than air lower in Earth's atmosphere.
2 Air near the middle of the atmosphere is denser than air higher above it.
3 The weight of all the air above presses down on the air particles closest to Earth, forcing them close together.
10000KG

✓Concept Check

1. How does temperature affect air pressure?

____

____

2. Fill in the chart. Check the correct column.

	Dense Air	**Less Dense Air**
Warm		
Cold		
Rises		
Sinks		
High pressure		
Low pressure		

✓Concept Check

1. The **Main Idea** on these two pages is Air is always moving and changing. **Details** tell more about the main idea. Circle two details that tell how air moves and changes.

2. Why is the water cooler than the sand at the beach?

____

3. Why is air over the sand hotter than air over the water?

____

Uneven Heating and Local Winds

How did the air feel when you left home this morning? Was it hot or cold? Was it windy or was it calm? Does the air feel the same right now? It probably does not feel the same. Air is always moving and changing.

The sun is always sending heat toward Earth. Some of it bounces off air and clouds. Earth absorbs the rest of the heat from the sun. However, land and water absorb heat in different ways.

Soil warms up faster in the sun than water does. That means at the beach, it is cooler in the water. During the day, the sand is hotter than the water. The sand gives off more heat. If the sand is hot, the air over the sand is hot, too.

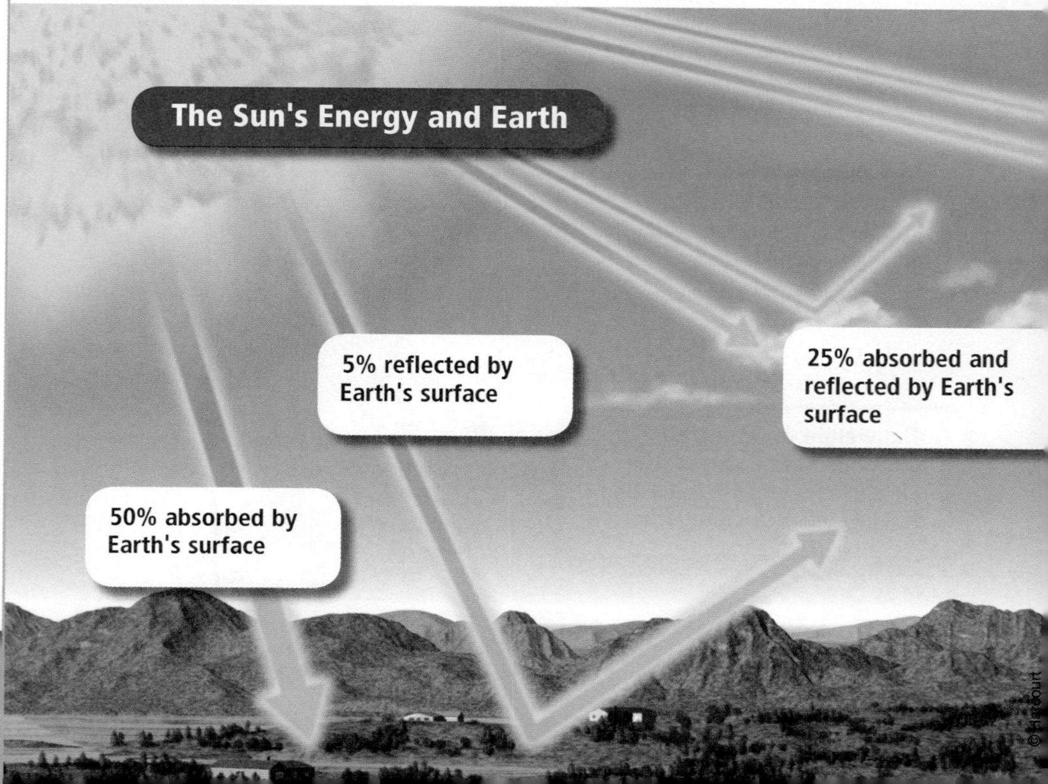
The Sun's Energy and Earth
5% reflected by Earth's surface
25% absorbed and reflected by Earth's surface
50% absorbed by Earth's surface

Remember water is cooler than the sand, during the day. That means the air over the water is cooler.

Cool air is denser than warm air. Cool air sinks. Warm air is less dense, so cooler, denser air pushes it up. This air movement in the atmosphere is called a **convection current**.

Air also moves horizontally. When cool air sinks it spreads out along the surface. Air moves from low pressure areas to high pressure areas. This moving air is called wind. Sometimes, areas close together have different temperatures. This is because of uneven heating. Different temperatures cause different air pressures. This makes **local wind**. Local winds often happen along shores, like the beach.

20% absorbed and reflected by air.

The arrows in this diagram show that in convection currents, air moves across the Earth's surface as well as up and down.

✓ Concept Check

1. Tell one detail about cool air.

____

2. Why do places close together sometimes have different temperatures?

____

3. In what directions does air move?

____

✓Concept Check

1. The **Main Idea** on these two pages is Global winds occur over large parts of Earth. **Details** tell more about the main idea. Circle two details about global winds.
2. The prevailing westerlies usually blow from west to east. Color all of the arrows on the drawing that show prevailing westerlies.
3. Look back at page 9. How are local and prevailing winds different?

____

____

____

____

Global Winds

A **prevailing wind** is a global wind. It occurs over a very large part of Earth. It almost always blows from the same direction.

Prevailing winds result from uneven heating of large areas on Earth. Local winds result from uneven heating of a small area.

In the United States, weather is caused mostly by *prevailing westerlies*. These are cooler winds moving south. Prevailing westerlies generally blow from west to east.

The prevailing westerlies help weather forecasters. Weather in California will probably head toward Kansas.

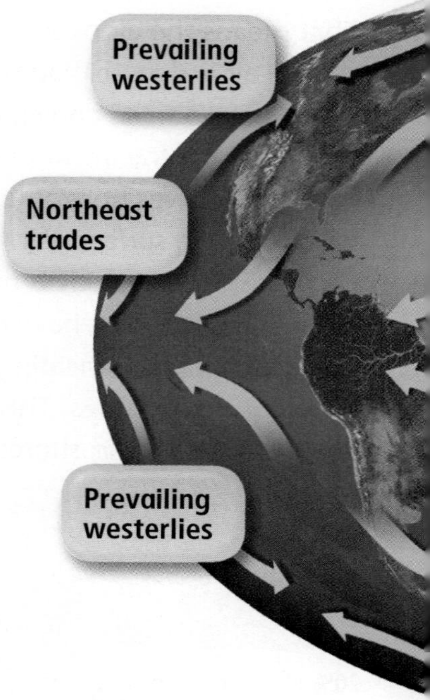
Prevailing westerlies
Northeast trades
Prevailing westerlies

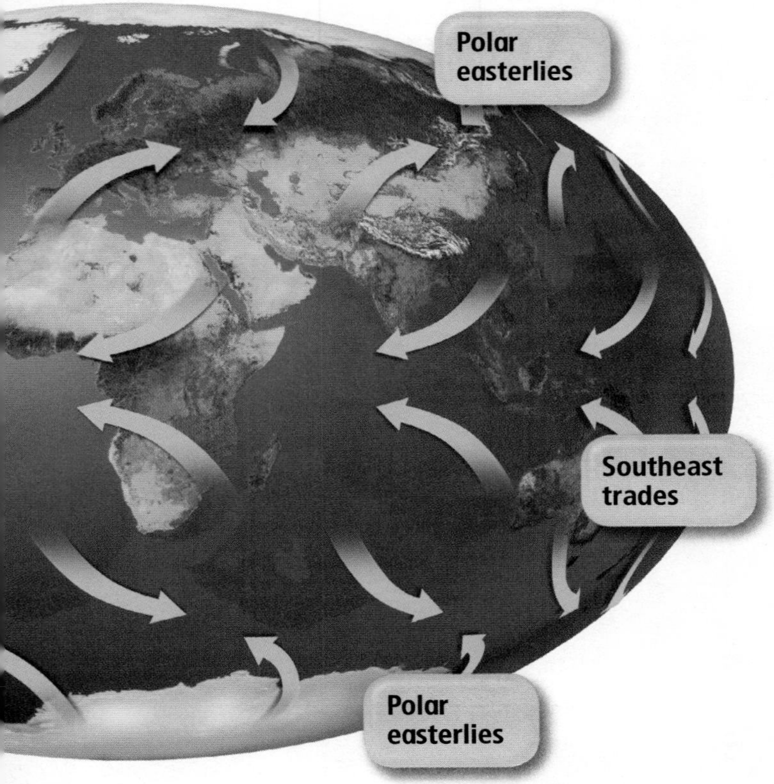
Polar
easterlies
Southeast
trades
Polar
easterlies

Lesson Review

Complete this Main Idea statement.

1. Uneven heating of Earth affects our ___________.

Complete these Detail statements.

2. The upward and downward movement of air is called a ___________ current in the atmosphere.
3. During a hot day, cooler air over the water is more ___________ than the warmer air over the land.
4. Local winds move a short distance. A ___________ wind is a global wind that moves great distances.

California Standards in This Lesson

4.b *Students know* the influence that the ocean has on the weather and the role that the water cycle plays in weather patterns.

Vocabulary Activity

Watery Words

This lesson is about the way the water cycle affects weather around the world.

1. Choose one of the vocabulary words. Write a sentence using the word.

____

____

____

2. Choose another vocabulary word. Draw a picture that shows the word's meaning.

Lesson 2

How Do the Oceans and the Water Cycle Affect Weather?

VOCABULARY
current
humidity
precipitation

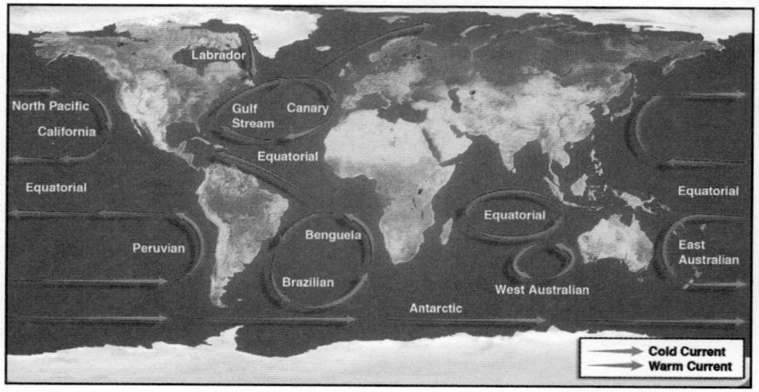

Labrador
North Pacific
California
Gulf Stream
Canary
Equatorial
Equatorial
Peruvian
Benguela
Brazilian
Antarctic
Equatorial
West Australian
Equatorial
East Australian
Cold Current
Warm Current

A **current** flows like a river through the ocean.

Active people sweat in high **humidity**.

Precipitation can fall any time of year.

Hands-On Activity Precipitation

Pour enough water into two cups to cover the bottoms of the cups. Seal both cups with plastic. Set them in the sun until you see the water on the plastic.

1. Place one cup in the freezer for half an hour. What do you see when you take it out?

____

2. Put the other cup under a bright lamp for half an hour. What happened to the water?

____

3. How do the cups model the water cycle?

____

____

____

✓Concept Check

1. A **Cause** is something that makes another thing happen. An **Effect** is the thing that happens. Underline the cause of air over the oceans being cooler in summer. Circle the effect of the California current moving north to south.

2. Fill in the missing label on the graph.

What Covers Earth's Surface?

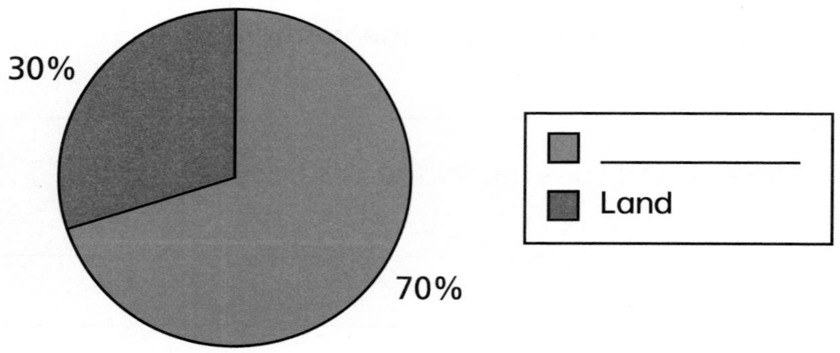
30%
70%
Land

3. Where does heat come from in winter?

____

4. Why don't ocean temperatures change much during the year?

____

The Oceans Affect Weather

The oceans cover about 70 percent of Earth's surface. You learned that land heats up and cools off faster than water. During summer, oceans absorb less heat from the sun than land. So the oceans and the air over them are cooler.

In winter, oceans are warmer than land. The oceans release heat into the air above them. This keeps the planet warmer. If Earth had no oceans, the temperatures would be more extreme.

Ocean temperatures do not change much during the year. The sun heats different parts of the oceans unevenly. Water on the ocean's surface is pushed forward by winds. The winds cause a **current**. An ocean current is a stream of water that flows like a river. Currents move heat over great distances through the ocean.

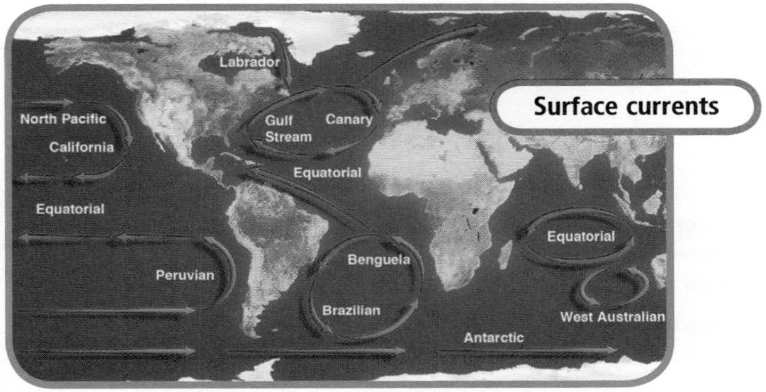
Surface currents
Labrador
North Pacific
California
Gulf Stream
Canary
Equatorial
Equatorial
Equatorial
Benguela
Peruvian
Brazilian
West Australian
Antarctic

The Gulf Stream current flows across the Atlantic Ocean. It begins in the tropics and moves northeast. This current carries warm water toward countries in northern Europe. The warm current causes the weather in those places to be warmer. Without this current, the weather would be much colder in those regions.

The California Current flows from north to south. It keeps the weather in California cooler.

In most years, wind pushes warm water away from the west coast of South America. The warm water evaporates fast into clouds. As a result, Australia gets warm water currents, clouds, and rain. The west coast of North America stays dry.

Some years, the winds reverse direction. The weather pattern also reverses. This is called an El Niño. Australia gets dry weather and western North America has wetter weather. The changing winds do not blow the warm surface water away from South America's coast. The South American coastal waters remain warm. The warm water makes clouds and storms on the U.S. west coast.

✓Concept Check

1. Look at the map. Circle the names of the currents that flow near California.

2. What causes the weather in northern Europe to get warmer?

____

____

3. Look at the map. Which current moves water away from the west coast of South America?

____

4. Fill in the chart. Compare a normal weather pattern to an El Niño.

	Normal	**El Niño**
Weather in Australia	clouds rain warm water	
Weather on west coast of North America	dry	

✓Concept Check

1. A **Cause** is something that makes another thing happen. An **Effect** is the thing that happens. Underline the cause of water vapor condensing. Circle an effect of water vapor condensing.

2. What happens to liquid water that falls to Earth?

____

____

3. Look at the drawing. Circle the step in which the sun is heating the water.

4. List three things that the sun's heat causes.

   1. ___________________________________
   2. ___________________________________
   3. ___________________________________

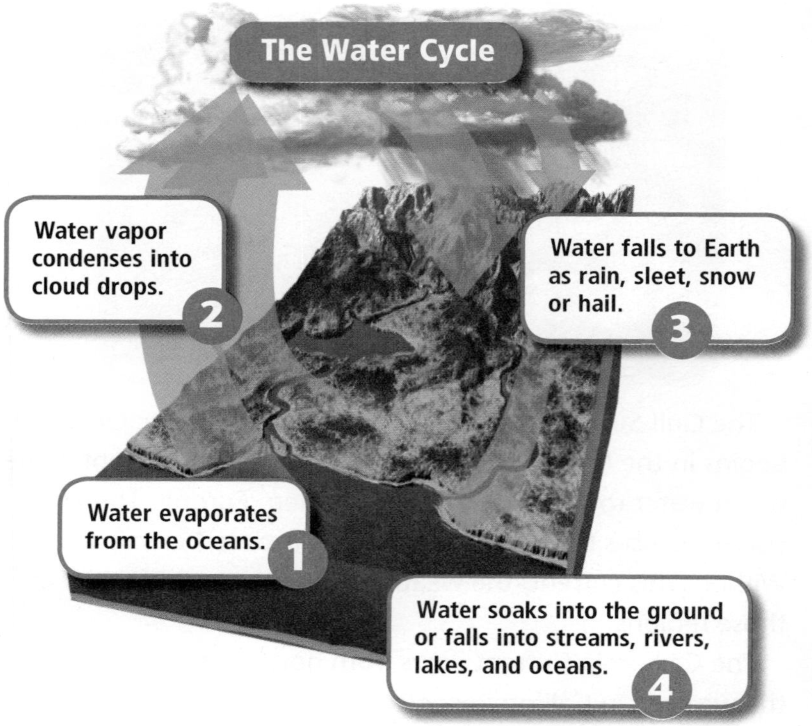
The Water Cycle
Water vapor condenses into cloud drops. 2
Water falls to Earth as rain, sleet, snow or hail. 3
Water evaporates from the oceans. 1
Water soaks into the ground or falls into streams, rivers, lakes, and oceans. 4

Weather Patterns and the Water Cycle

Water is constantly moving and changing states. It moves from Earth's surface to the atmosphere and back to Earth.

The sun's heat causes convection currents in the atmosphere. It affects currents in the oceans. The sun's heat also causes the movement of the water cycle.

During the water cycle, liquid water is heated by the sun and evaporates. It turns into a gas called water vapor. Water vapor stays a gas as long as it is warm. When water vapor cools, much of it condenses back into a liquid form. The liquid water falls back to Earth. Some of the water soaks into the ground. Some of it runs into streams, rivers, lakes, and oceans. The water cycle starts again as the sun heats the ocean waters.

Heat Transfer Through the Water Cycle

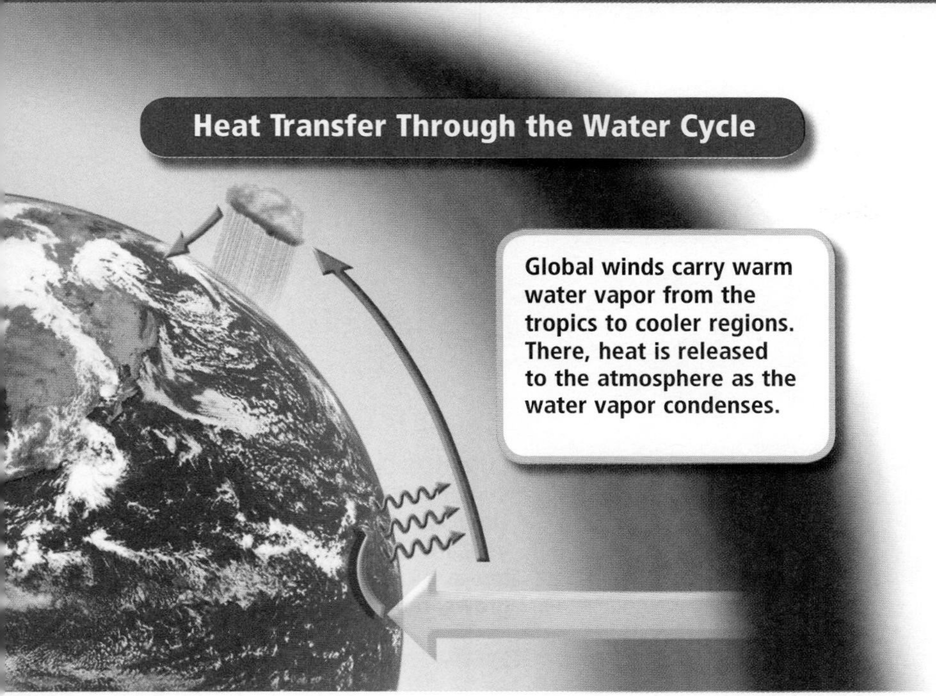
Global winds carry warm water vapor from the tropics to cooler regions. There, heat is released to the atmosphere as the water vapor condenses.

Air in the atmosphere moves in convection currents. Water evaporates from the oceans near the warm tropics. The evaporated water moves long distances on global winds. If there were no global winds, what would happen? Almost all Earth's rain would fall near the equator.

Warm air flows from the equator north and south toward the poles. The warm air carries water vapor. Water vapor carried by global winds contains heat. As the water vapor cools, the vapor condenses back into liquid water. The condensation releases heat energy into the atmosphere. Both heat and water move through the water cycle.

This process helps balance temperatures in the atmosphere. The tropics lose some heat and water vapor. The cooler regions gain heat and moisture.

✓Concept Check

1. In the picture, circle the symbol that shows warm air carrying water vapor rising into the atmosphere. Put a box around the symbol that shows condensed water vapor falling to Earth as rain.

2. When is heat released into the atmosphere by the water cycle?

   ____

3. How does the water cycle affect weather?

   ____

4. How does air in the atmosphere move?

   ____

✓Concept Check

1. A **Cause** is something that makes another thing happen. An **Effect** is the thing that happens. Underline the cause of clouds forming. Circle the effect of many clouds forming.

2. What causes high humidity?

____

3. What causes snow to fall?

____

4. Draw each type of precipitation in the space below.

Rain	
Snow	
Sleet	
Hail	
Fog	

CUMULUS CLOUDS
Cumulus (KYOO•myuh•luhs) clouds are puffy. They indicate fair weather, but as a cumulus cloud grows, rain can develop.

Clouds

Weather is part of the water cycle. Water enters the atmosphere when it evaporates. The amount of water vapor in the air is **humidity**. A lot of water vapor in the air is high humidity. Very little water vapor in the air is low humidity.

Humidity also depends on the air's temperature. Warm air can have more water vapor in it. Cold air has less water vapor in it.

Air that has high humidity creates clouds. As warm air is forced up, it cools, Some of the water vapor condenses on dust in the air. As more and more water condenses, a cloud forms.

The water in clouds returns to Earth as rain, snow, or sleet. Water that falls from the atmosphere is **precipitation**.

Precipitation

Precipitation falls as rain, snow, sleet, or hail. Snow forms when water vapor turns directly into ice crystals. Sleet and hail form when liquid water passes through air that is cold enough to freeze water drops.

Fog is a similar weather condition. Fog is water vapor that condenses into small water droplets near the ground.

Lesson Review

Complete these Cause and Effect statements.

1. In the winter, the oceans release heat into the air, and this makes the planet ____________.
2. Australia has drier weather and western North America has wetter weather than normal during an ____________ ____________.
3. When water droplets in clouds get too heavy, water falls as ____________.
4. Both ____________ and water are moved through the water cycle.

California Standards in This Lesson

4.d *Students know* how to use weather maps and data to predict local weather, and that weather forecasts depend on many variables.

Vocabulary Activity

Predicting Weather

In this lesson, you will learn how weather is predicted. You will also learn about how people study the weather.

1. The suffix *-ology* means "the study of." The suffix *–meter* means "instrument for measuring." Write four of the vocabulary terms in the correct places on the chart.

Words That Name Something to Study	Words That Name an Instrument

2. Look at the terms *air mass* and *front.* Tell how you have heard the terms *air mass* and *front* used before.

____

____

____

Lesson 3

How Is Weather Predicted?

VOCABULARY

meteorology
barometer
anemometer
hygrometer
air mass
front

Scientists who study **meteorology** use weather stations and maps.

Stormy
RAIN
Change
FAIR
Very Dry

A **barometer** can help people decide what to wear. Low pressure means a chance of rain.

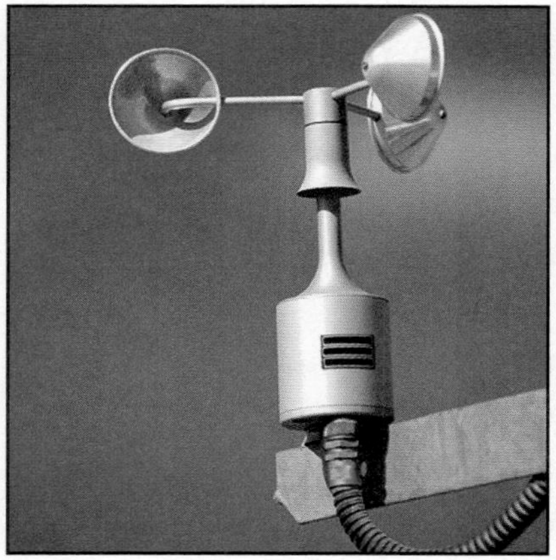

This **anemometer** can tell how fast the wind is blowing.

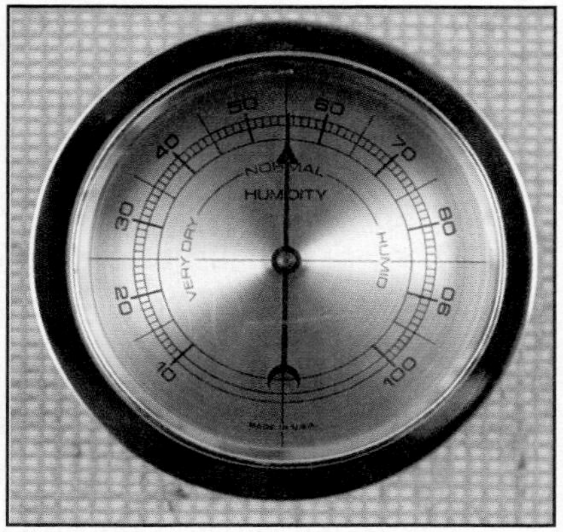

NORMAL
HUMIDITY
VERY DRY
HUMID
10
20
30
40
50
60
70
80
90
100

People in tropical climates might use a **hygrometer** to measure the humidity.

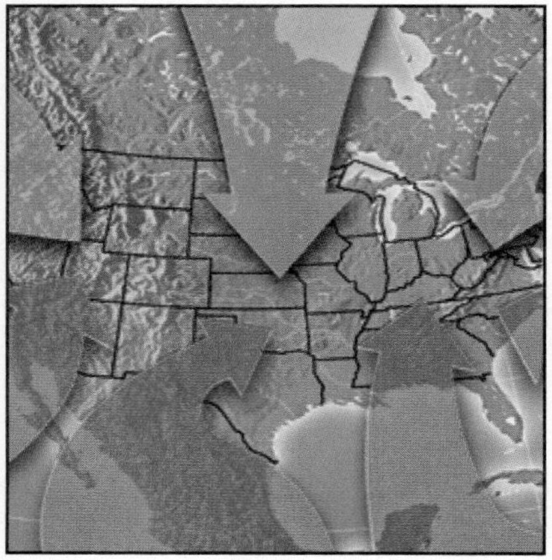

A wet, tropical **air mass** brings warm, humid weather.

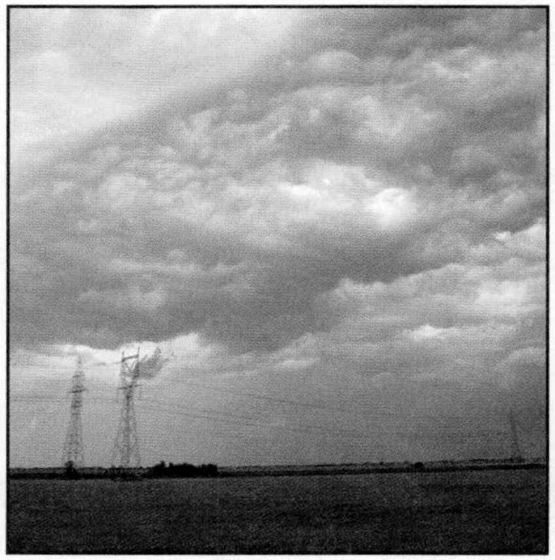

Sometimes, you can locate a **front** by watching the clouds.

Hands-On Activity
Weather Forecast

Look at a picture of a weather map in a newspaper. Using the symbols, make a weather forecast for tomorrow. Present your forecast to the class as a meteorologist would.

1. What did you base your predictions on?

____

2. The next day, check the weather for your area. Were your predictions correct?

____

3. How can you use local weather maps?

____

____

✓Concept Check

1. The **Main Idea** on these two pages is Weather can be measured. **Details** tell more about the main idea. Underline two details that tell how weather can be measured.
2. What is a meteorologist?

___

3. If the barometer rises, how will the weather change?

___

4. If it is raining, will the barometer be rising or falling?

___

Measuring Weather

The study of weather is **meteorology**. A scientist who studies weather is a *meteorologist*. A meteorologist uses many different instruments to measure the weather. Four of these are thermometer, barometer, anemometer, and hygrometer. They can be used to predict weather. People who predict weather are making a *forecast*.

A **barometer** measures air pressure. When a barometer is rising, air pressure is increasing. Higher air pressure means cooler weather. Cold air usually has less water vapor than warm air. A rising barometer means less humidity. Less humidity means a lower chance of rain. When a barometer is rising, you can forecast cool weather with a low chance of rain. A falling barometer reading often means more humidity is in the air. More humidity means a higher chance of rain.

When making weather predictions, wind must be considered. **Anemometers** measure wind speed. *Windsocks* and *wind vanes* measure wind direction. Changes in wind speed and wind direction tell more about the weather. Say it is winter, and the wind begins to blow from the south. You can predict that the weather will become warmer.

Weather Instrument		Measures
	**Thermometer**	**A thermometer measures air temperature. If the air cools down during the day or warms up in the evening, the change is a sign that rain may fall soon.**
	**Barometer**	**A barometer measures air pressure. This is also called *barometric pressure.* A rapid change in air pressure often means the weather is about to change.**
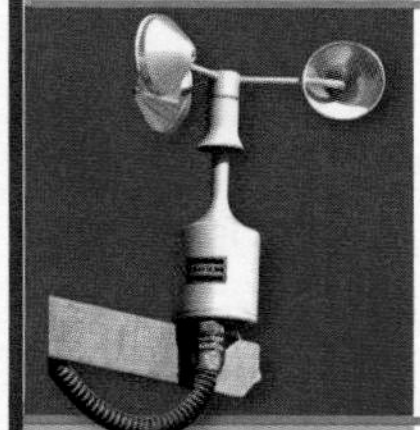	**Anemometer**	**An anemometer measures wind speed. Like a change in air pressure, a change in wind speed may mean that the weather is about to change.**
 0 20 40 60 80 100 120 -20 -40 -60	**Hygrometer**	**A hygrometer measures humidity. An increase in humidity often means it is about to rain.**

✓Concept Check

1. What kinds of changes can help predict weather?

____

2. Fill in the chart. Tell what instrument you would use to help you find out what you want to know.

What I Want to Know	Instrument I Would Use
the temperature	
whether or not it will rain	
how fast the wind is moving	
air pressure	

3. How does rising air pressure help people predict the weather?

____

____

____

4. What prediction can you make if it is winter and the wind starts blowing from the south?

____

____

____

✓ Concept Check

1. The **Main Idea** on these two pages is Air masses affect weather. **Details** tell more about the main idea. Underline two details that tell how air masses affect weather.

2. Tell which kind of air mass brings each kind of weather shown in the chart.

Kind of Weather	Air Mass

A *continental* air mass forms over land, so it is dry. (c)
A *maritime* air mass forms over water, so it is humid. (m)
A *polar* air mass forms over a cold area, so it is cold. (P)
A *tropical* air mass forms over a warm area, so it is warm. (T)

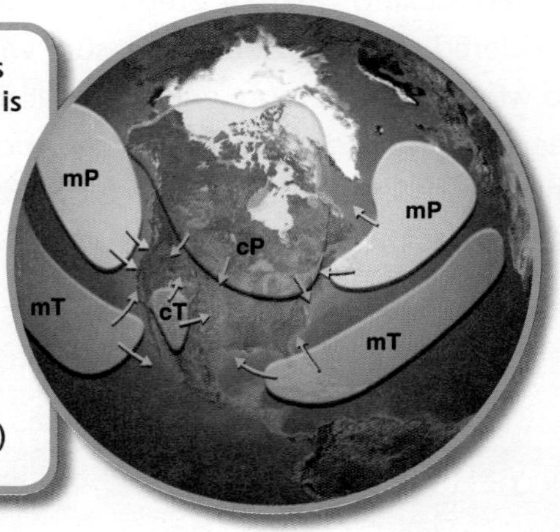
mP
mP
cP
mT
cT
mT

Air Masses

The sun heats Earth's atmosphere, land, and water. Some air is warm and some air is cool. The uneven heating causes the air to move. Air moves in a large, regular mass. An **air mass** is a large body of air that has the same temperature and humidity.

An air mass is similar to the region where it forms. An air mass that forms over the ocean near the equator is humid and warm. An air mass that forms over northern Canada is dry and cold.

Four kinds of air masses affect weather in our country. *Continental polar* air masses bring cool, dry weather. *Continental tropical* air masses bring hot, dry weather. *Maritime polar* air masses bring cold, humid weather. *Maritime tropical* air masses bring warm, humid weather. When weather changes in a region, the air mass changes too. A changing air mass produces winds. Air masses usually move from west to east across the United States.

Fronts

The border where two air masses meet is a **front**. Most weather changes occur along fronts.

The two main kinds of fronts are cold fronts and warm fronts. A cold front forms when a cold air mass moves under a warm air mass. The warm air mass is less dense. It is forced up quickly and begins to cool. When a warm, moist air mass cools, its water vapor condenses. This fast cooling and condensation causes heavy rain, thunderstorms, or snow. Cold fronts usually move quickly, so the storms do not last long.

A warm front develops differently. A warm air mass moves behind and then over a cold air mass. As the warm air slowly slides up and over the cold air, clouds form ahead of the front. The clouds produce rain or snow that can last for hours.

After a cold front passes through an area, the weather is cooler and drier.

✓ Concept Check

1. What occurs along a front?

____

2. Circle the name of the denser air mass.

   cold front **warm front**

3. How does a warm front form?

____

____

4. What kind of front brings heavy rains and thunderstorms that do not last long?

____

✓Concept Check

1. The **Main Idea** on these two pages is Weather maps can be used to help people learn about the weather in an area. **Details** tell more about the main idea. Underline two details about weather maps.

2. Look at the map. Circle the cold fronts.

3. Fill in the chart. Draw the symbol used to show the weather described.

Kind of Weather	Symbol
Warm front moving east	
Rainy weather	
A cold front moving south	

Weather Maps

People use weather maps to learn about the weather in an area. Weather maps use symbols to show different types of weather. A sun symbol means sunny weather. The symbol of a cloud with rain means rainy weather.

Fronts are shown on weather maps. The symbol for a warm front is a red line with half circles. A blue line with triangles shows a cold front.

If a warm front is moving east, the half circles are on the right side of the line. This shows the direction the front is moving. If a warm front is moving south, the triangles are on the bottom of the blue line.

Temperature is given on a weather map. Temperatures can be shown by a special color code on the map. Weather maps might also show wind speed and wind direction.

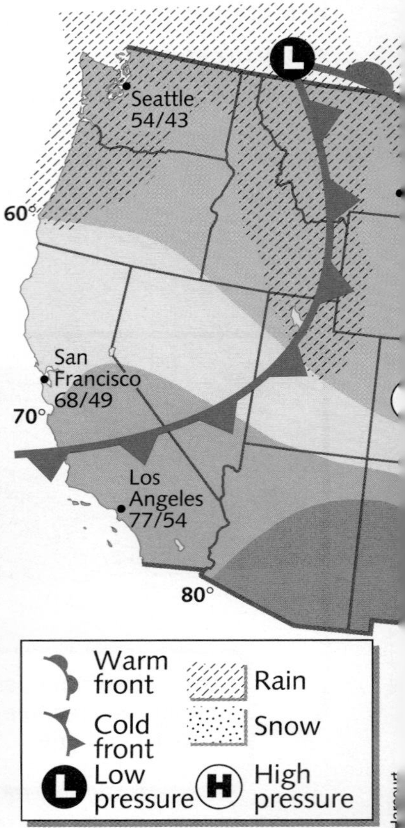

Seattle 54/43
60°
San Francisco 68/49
70°
Los Angeles 77/54
80°
Warm front
Cold front
Low pressure
Rain
Snow
High pressure

High and low-pressure systems may be indicated on a weather map. A high-pressure system is noted with an *H*. A high-pressure system is an area of cool, dense air. It is surrounded on all sides by warmer low-pressure air.

A low-pressure system is noted with an *L*. A low-pressure system forms when an area of warm, less-dense air is surrounded by cooler, high-pressure air.

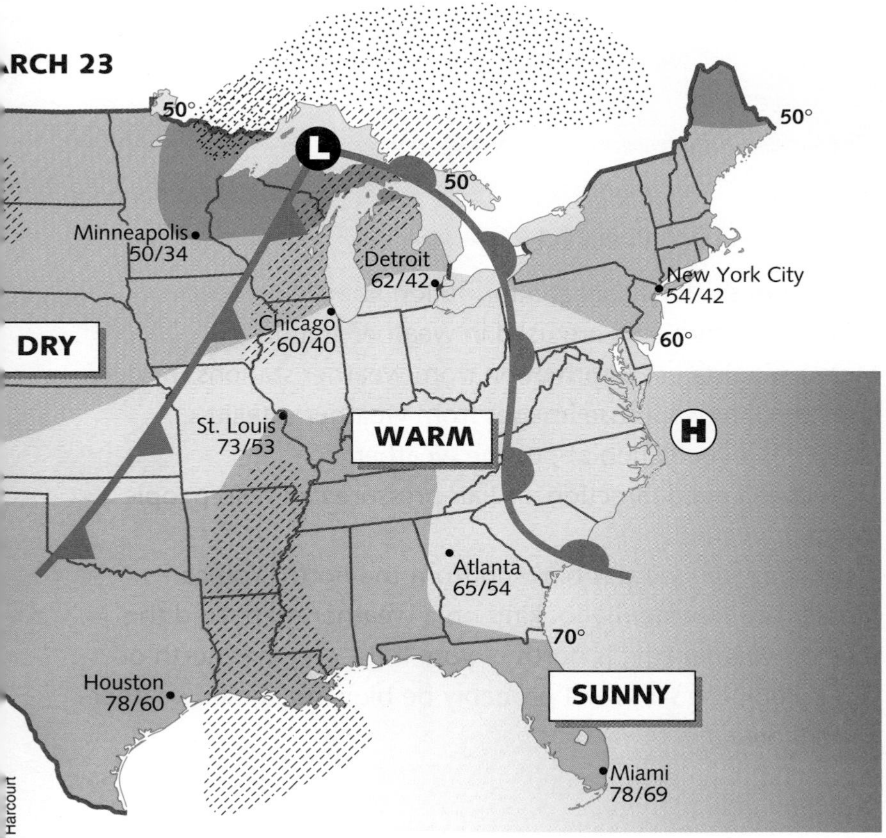

RCH 23
50°
50°
50°
L
Minneapolis
50/34
Detroit
62/42
New York City
54/42
DRY
Chicago
60/40
60°
St. Louis
73/53
WARM
H
Atlanta
65/54
70°
Houston
78/60
SUNNY
Miami
78/69

✓ Concept Check

1. Pretend you are a meteorologist. You need to give the weather report. Use the weather map to help. Give the weather report for your own area first.

My Weather Report

It is between ___________ and ___________ degrees in ___________.

A ___________ front is moving east.

Seattle is having ___________ weather.

There is ___________ pressure over Denver.

The hottest city in the country is ___________.

It is 58 degrees in ___________.

2. Fill in the chart. Tell what the symbols mean.

Symbol	Meaning
L	
H	

3. What kind of information can you find on a weather map?

____

____

✓Concept Check

1. The **Main Idea** on these two pages is Weather can be predicted. **Details** tell more about the main idea. Underline two details that tell how weather can be predicted.

2. If the weather north of you is cold, what kind of weather will you have over the next few days?

____

3. How can air pressure help predict weather?

____

____

4. What does a meteorologist need to predict the weather?

____

____

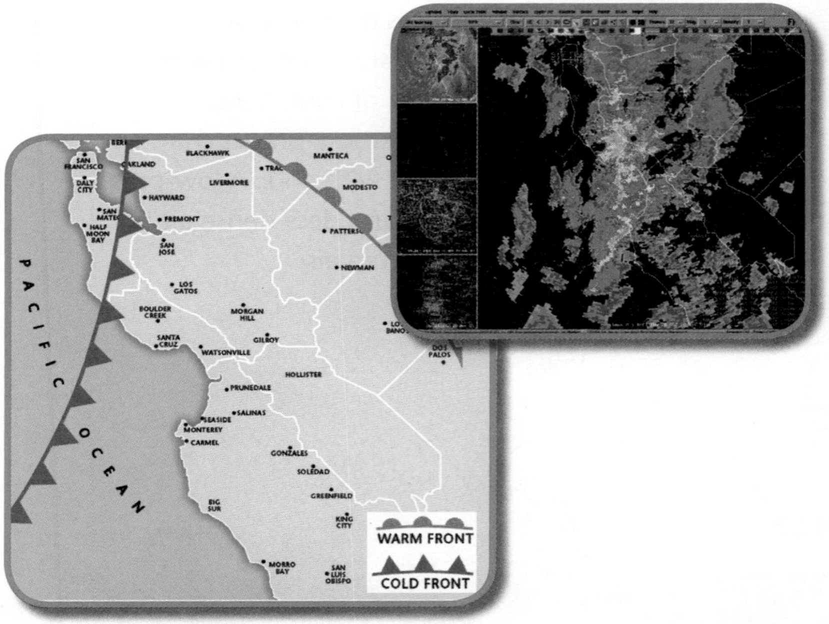

PACIFIC OCEAN
WARM FRONT
COLD FRONT

Forecasting Weather

Weather stations collect information about the weather. Different instruments are used in weather stations. Meteorologists use information from weather stations to make forecasts. They also use images from weather satellites. A *forecast* is a prediction about the weather.

Studying wind direction and air pressure can help people forecast weather.

Let's say that wind is blowing from the north. You can predict the weather by looking on a weather map. Find the type of weather that is north of you. If the weather north of you is cool, the wind will probably be blowing cooler weather to your area.

Air pressure is another way to predict weather. A low-pressure system on a weather map forecasts stormy weather. A high-pressure system on the weather map means the weather will probably be fair.

Weather predictions for the next few days are usually accurate. It is difficult to predict weather farther into the future. Small changes in temperature and air pressure can cause big changes in weather. Scientists cannot accurately predict the weather more than a few days ahead.

Lesson Review

Complete this Main Idea statement.

1. Weather maps can help you predict, or ___________ the weather for the near future.

Complete these Detail statements.

2. Air pressure is measure with a ___________.
3. A continental tropical ___________ ___________ produces hot, dry weather.
4. Most weather changes occur along ___________.

California Standards in This Lesson

4.c *Students know* the causes and effects of different types of severe weather.

Vocabulary Activity

Severe Weather

In this lesson, you will learn why severe weather occurs.

1. Tell about a thunderstorm you have seen.

____

____

____

2. Look at the pictures. Answer the following questions with a vocabulary word.

Which type of weather can pull trees out of the ground? ____________

Which type of weather brings heavy rains?

____

Which type of weather can cause a great deal of damage? ____________

Lesson 4

What Are the Causes and Effects of Severe Weather?

VOCABULARY

monsoon
hurricane
thunderstorm
tornado

A **monsoon** can bring heavy rains.

I am not afraid of **thunderstorms**. I like to hear the rain and thunder. The lightning is exciting.

A **tornado** can cause a great deal of damage.

A **hurricane** can pull trees out of the ground.

Hands-On Activity
Bottle Tornado

Fill one 2-liter plastic soda bottle with water. Stand by a sink and turn the bottle upside down. Immediately begin to swirl the water by moving the bottle hard in a clockwise motion. Keep swirling while the water pours out.

1. Did you see the tornado shape in the bottle? Did the water pour out quickly?

____

____

2. Repeat the experiment, but this time do not swirl the water. Did the water flow out faster or slower?

____

3. In a tornado, the humid air spins up. How did this activity show how a tornado's air spins?

____

✓Concept Check

1. A **Cause** is something that makes another thing happen. An **Effect** is the thing that happens. Underline the cause of California's monsoon season. Circle an effect of the monsoon season in California.

2. What are the Santa Ana winds?

____

3. Fill in the chart. Check which phase of the monsoon system in California has the weather described.

	Dry	Wet
Santa Ana winds blow in		
Pacific waters are cooler than land		
Cool and cloudy		
Dry and warm		
Winds come from land		

Pacific Storms

In North America, weather usually moves from west to east. Because the Pacific Ocean is west of California, the ocean affects California's weather. California has a monsoon pattern because of the Pacific. A **monsoon** is a wind system that reverses its direction with the seasons.

When the land cools in the winter, the air above it cools too. The cooler dense air over land moves toward the Pacific. It forces the ocean's warmer air to rise. Because the wind comes from the land, it is dry. The Santa Ana winds are part of the dry phase of southern California's monsoon.

During the summer, the opposite effect happens. The Pacific waters are cooler than the land. The cool air from the Pacific is humid. When the humid air reaches the warm land, the air warms up. The air becomes less dense and is forced upward. As the humid, warm air rises, it cools and clouds form.

A cyclone is another type of Pacific weather pattern that affects California. A *cyclone* is an air mass that is turning rapidly.

Because of the Earth's rotation wind does not blow in straight lines. The air curves as it blows. The warm less-dense air rotates and rises over the cooler low-pressure area. A circular wind storm forms.

Winter storms come in from the Pacific Ocean. They can bring heavy rains to the coast and heavy snows to the mountains.

✓ Concept Check

1. What is a cyclone?

____

2. Why doesn't wind blow in straight lines?

____

3. Where do California's winter storms come from?

____

4. What is one reason it rains during the summer in California?

____

✓ Concept Check

1. A **Cause** is something that makes another thing happen. An **Effect** is the thing that happens. Underline the cause of tropical storms turning into hurricanes. Circle an effect of hurricanes.

2. Complete the chart of hurricane facts.

Wind speeds	
Movement of wind	
Kind of air in the eye	
Most intense part of the storm	
Height of clouds	

3. What is the difference between a tropical depression and a hurricane?

____

Hurricanes and Other Cyclones

A type of cyclone that begins over warm ocean water is a hurricane. A **hurricane** is a large rotating tropical storm. It has wind speeds of at least 119 km/hr (74 mi/h).

A hurricane is first called a *tropical depression*. This is because the air pressure is low, or depressed. Winds rotate around the low-pressure center of the tropical depression. When the winds reach 63 km/hr (39 mi/h), the tropical depression becomes a *tropical storm*. A tropical storm is given a name. As the storm grows stronger, it becomes a hurricane.

The eye of a hurricane is a calm center. It is dry, cool air that the storm pulls down. Around the eye is the eye wall. The eye wall is the most intense part of the storm.

Anatomy of a Hurricane

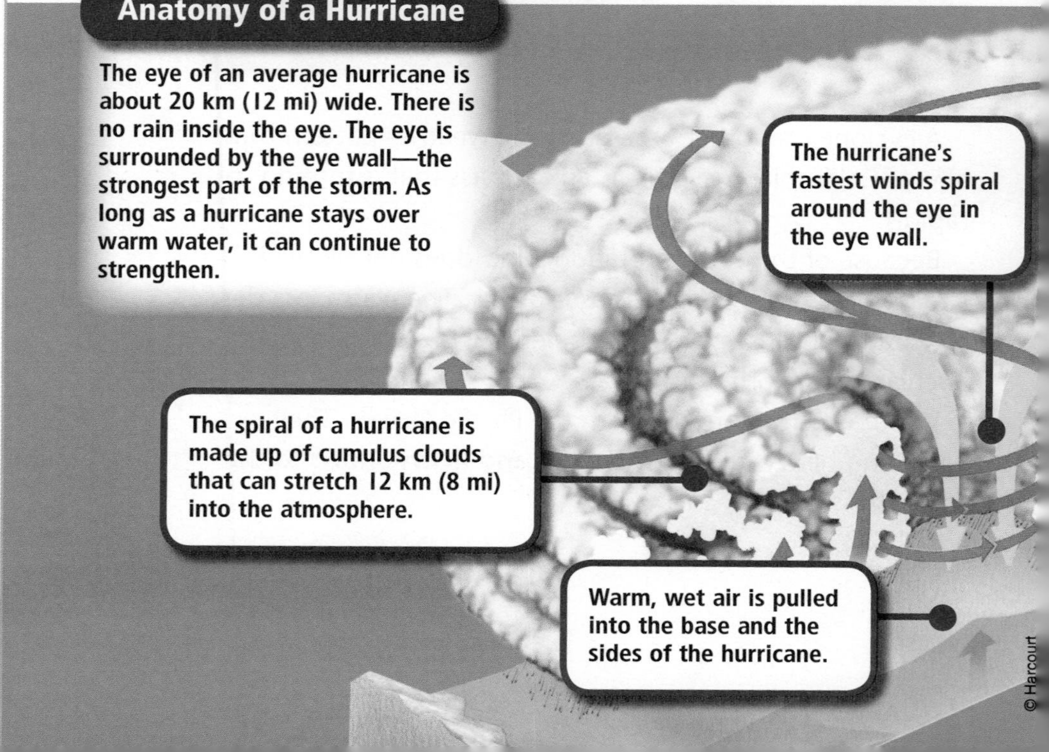
The eye of an average hurricane is about 20 km (12 mi) wide. There is no rain inside the eye. The eye is surrounded by the eye wall—the strongest part of the storm. As long as a hurricane stays over warm water, it can continue to strengthen.
The hurricane's fastest winds spiral around the eye in the eye wall.
The spiral of a hurricane is made up of cumulus clouds that can stretch 12 km (8 mi) into the atmosphere.
Warm, wet air is pulled into the base and the sides of the hurricane.

Water vapor that is carried up condenses into rain. Condensation releases heat. Heat and moisture increase the energy of the storm.

The storm pushes the ocean's surface. This causes the ocean surface to rise. These rises are *storm surges*. Other hurricane dangers include high winds, heavy rains, flooding, and tornadoes.

The word *hurricane* is not used everywhere. In the northern Pacific Ocean, near Japan or China, this kind of storm is called a *typhoon*.

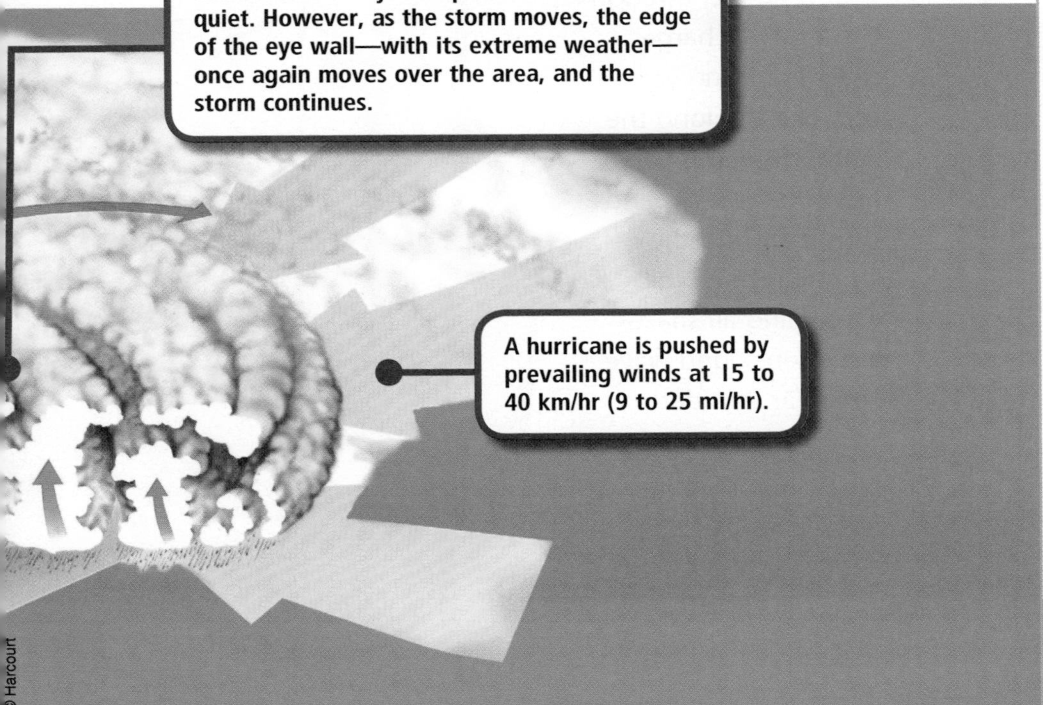

The hurricane's eye is a place of calm and quiet. However, as the storm moves, the edge of the eye wall—with its extreme weather—once again moves over the area, and the storm continues.

A hurricane is pushed by prevailing winds at 15 to 40 km/hr (9 to 25 mi/hr).

✓Concept Check

1. What can cause a hurricane to get stronger?

____

2. Look at the drawing. Circle the hurricane's eye.

3. What is another word for *hurricane*?

____

4. Page 151 lists five dangers of hurricanes. Sort them using the chart below.

Wind Dangers	Water Dangers

✓Concept Check

1. A **Cause** is something that makes another thing happen. An **Effect** is the thing that happens. Underline the cause of thunderstorms. Circle an effect of winds spinning in a column of air.

2. What is lightning?

____

3. What sound does expanding air around a lightning bolt cause?

____

4. Fill in the chart. Tell whether each cause has the effect of a tornado or a thunderstorm.

Cause	Effect
Warm humid air moves upward quickly.	
Winds spin in a column of air.	
Warm, humid air is pulled into a funnel-shaped column.	
Electric charges build up in the bottom of a cloud.	
Warm air cools, rises, condenses and forms clouds.	
Low pressure makes nearby air rush in.	

Thunderstorms and Tornadoes

A **thunderstorm** is a storm with rain, wind, lightning, and thunder. Sometimes, a thunderstorm produces hail. About 45,000 thunderstorms occur on Earth every day! A thunderstorm begins to form when warm, humid air moves upward quickly.

The sun heats an area on Earth's surface. This creates a warm air mass above the land. A cold front may push under the warm air mass, or wind may force it up.

The warm air cools as it rises. The water vapor in it condenses and forms clouds. Soon rain begins to fall. The falling rain pulls cool air down with it. Winds blow both upward and downward in the cloud.

Electric charges build up in the bottom of the cloud. The charges travel through the air. The charge is the lightning you see. The air along the lightning bolt is very hot. The heat expands the air around the lightning bolt. The expanding air makes a sudden sound wave. You hear thunder!

Whenever meteorologists discover a strong thunderstorm, they look for something else. They look for another severe weather event called a tornado. A **tornado** is a violently rotating column of air.

A tornado forms when winds spin in a column of air. The column grows out of the bottom of a cloud. Warm, humid air is pulled into the funnel-shaped column. The humid air spins up and curves around the center. Low pressure in the center of the funnel makes nearby air rush in.

The path of a tornado is usually very narrow. It can last from a few minutes to a few hours. Tornadoes are more difficult to predict than other storms. Meteorologists issue warnings so people have time to take cover.

Lesson Review

Complete these Cause and Effect statements.

1. The ____________ Ocean affects much of California's weather.
2. In the summer, a ____________ will cause a layer of clouds to hang over the coast of California.
3. The eye of a ____________ is a calm center, caused by dry, cool air that is pulled down.
4. A ____________ forms when winds spin in a column of air that grows out of the bottom of a cloud.

Circle the letter in front of the best choice.

1. What happens to warm, less dense air?

   A It sinks and cools.
   B It rises and cools.
   C It rises and warms.
   D It sinks and warms.

2. Which is TRUE of the water cycle?

   A It takes in water from outside the atmosphere.
   B It releases water outside the atmosphere.
   C It affects weather around the world.
   D Temperature does not affect it.

3. How does the sun heat Earth?

   A evenly all around the world
   B unevenly over land and water
   C by conduction
   D through the water cycle

Use the picture to answer questions 4 and 5.

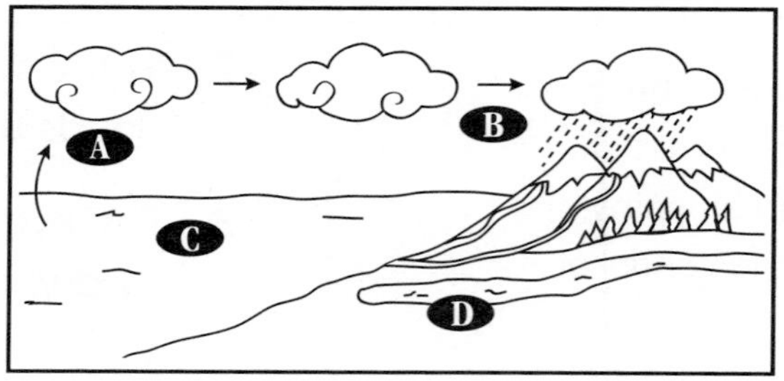
A
B
C
D

4. What does the picture show?

   A the sun's energy being absorbed
   B a cold front moving in
   C the water cycle
   D a warm front moving in

5. What is happening at step B?

   A precipitation
   B condensation
   C evaporation
   D dryness

6. Which statement is NOT true about the ocean?

   A Water evaporates from it during the water cycle.
   B It has warm and cold currents.
   C It does not affect Earth's weather.
   D In winter, it is warmer than land.

7. What causes local winds?

   A uneven heating of Earth
   B the water cycle
   C the sun's energy
   D convection currents

8. Which winds help weather forecasters in California?

A Polar easterlies
B Prevailing westerlies
C Northeast trades
D Southeast trades

9. What causes air to move?

A uneven heating of air masses
B being from certain regions
C type of air mass
D a front forming

10. What is unique about a monsoon system?

A It moves air.
B It affects weather.
C It reverses direction.
D It moves from west to east all the time.

11. In which direction or directions does air pressure push? Draw arrows around the word *air* to show.

air

12. What is weather?

____

____

____

____

13. What tools can help a meteorologist predict weather?

____

____

____

____

14. Look back to the question you wrote on page 116. Do you have an answer for your question? Tell what you learned that helps you understand weather.

____

____

____

____

____

____

The Solar System

In this unit, you'll learn about what lies beyond Earth's atmosphere—in space.

Thinking Ahead

What do you think the sun is? What do you think it is made of?

____

____

____

Do you know any other planets besides Earth? Draw one here.

Earth is round. What do you think keeps people from falling off the Earth as it rotates?

____

____

Write a question you have about the solar system.

____

____

____

Recording What You Learn

On this page, record what you learn as you read the unit.

Lesson 1

What is the sun made of?

____

____

How big is our sun in comparison to other stars?

____

Lesson 2

Draw the solar system and label the planets. Don't forget the asteroid belt!

Lesson 3

How do gravity and inertia keep planets from falling out of their orbits?

____

____

____

____

California Standards in This Lesson

5.a *Students know* the Sun, an average star, is the central and largest body in the solar system and is composed primarily of hydrogen and helium.

Vocabulary Activity

The Sun

In this lesson, you'll learn all about the sun. You'll find out what it's made of and why it has so much energy.

1. The chart below shows two of your vocabulary words. Write each word as part of a compound word. How many compound words can you write for each vocabulary word?

Vocabulary	Compound Words
sun	
star	

2. What is the base word of *fusion*?

Lesson 1

What Is the Sun?

VOCABULARY
star
sun
fusion

A **star** looks like a point of light in the night sky.

The **sun** is a star. We can see it during the day.

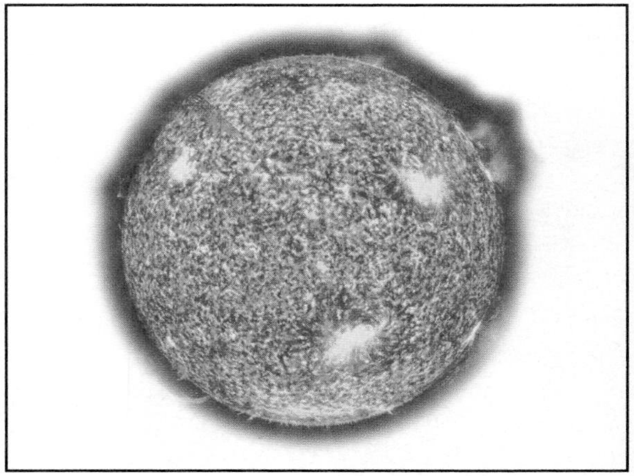

Most of the sun's energy comes from **fusion**. Two hydrogen atoms fuse and make a helium atom.

Hands-On Activity
Sun, Moon, and Stars

1. Draw a picture of our sun. Remember that it is of average size.

2. Now imagine what other stars look like in comparison. Draw at least five stars using different colors and sizes.

3. Cut out your stars. Paste or tape them onto a piece of black paper to make a constellation.

4. Name your constellation. Then explain why you named it that way.

My constellation's name is ______________________.

____

____

____

____

✓Concept Check

1. The **Main Idea** on these two pages is The sun is a star. **Details** tell more about the main idea. Underline two details about the sun's features.

2. Tell one detail about the sun.

____

____

3. Scientists classify stars by size, temperature, and color. Classify the sun.

Characteristic	Classify
Size	
Temperature	
Color	

4. What is the largest body in our solar system?

____

Stars

A **star** is a huge ball of very hot gases. The **sun** is the star at the center of our solar system. From Earth, the sun looks like a large ball of light.

The sun is the source of most energy on Earth. Without the sun, life on Earth could not exist.

Scientists classify stars by size, temperature, and color. The sun is an average size star. It is the largest body in the solar system. The sun has a medium temperature.

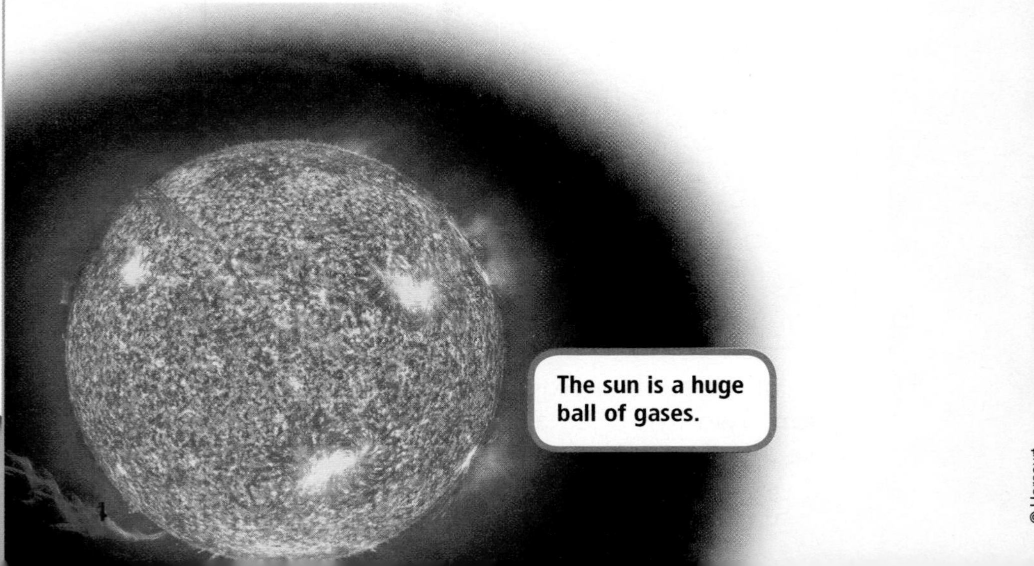

The sun is a huge ball of gases.

Features of the Sun

The sun has several layers. At the center of the sun is the core. Most of the sun's mass is in the core. The other layers are the *radiation zone*, the *convection zone*, and the *photosphere*. The photosphere is the surface of the sun. Above the photosphere is the sun's atmosphere.

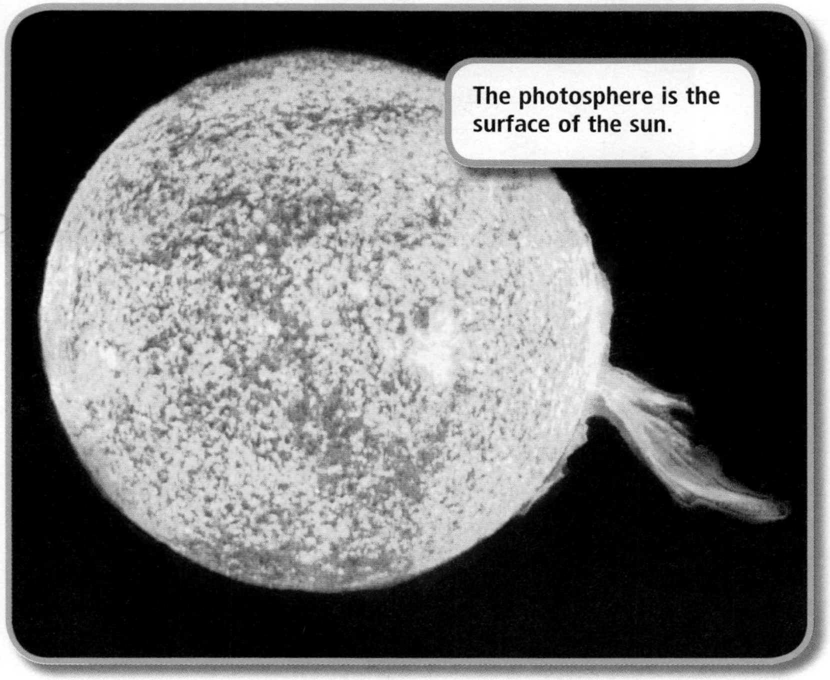

The photosphere is the surface of the sun.

✓ Concept Check

1. Name one layer of the sun.

____

____

2. Where is most of the sun's mass?

____

3. Look at the picture of the sun in the reader. What layer are you looking at?

____

Concept Check

1. The **Main Idea** on these two pages is The sun produces energy. Underline two details that tell how the sun produces energy.
2. What is the temperature at the center of the sun?

____

3. What are two of the gases that make up the sun?

____

4. Look at the drawing below. Fill in the missing labels to show where each layer of the sun is.

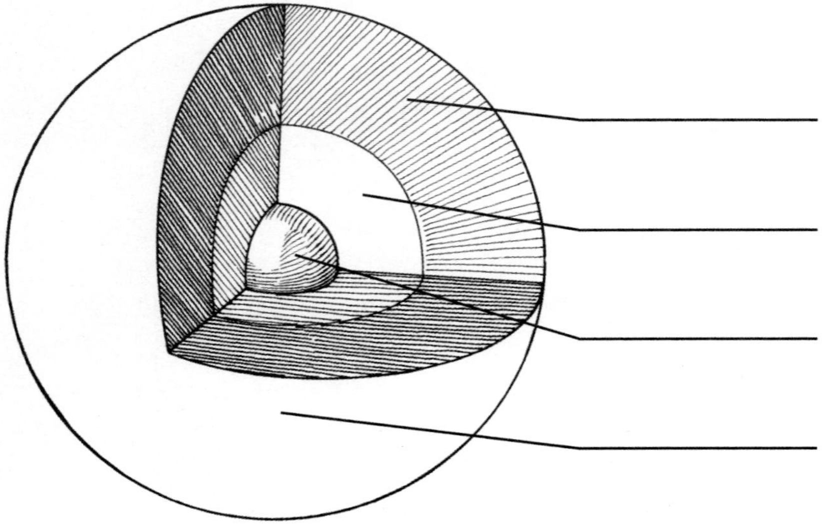

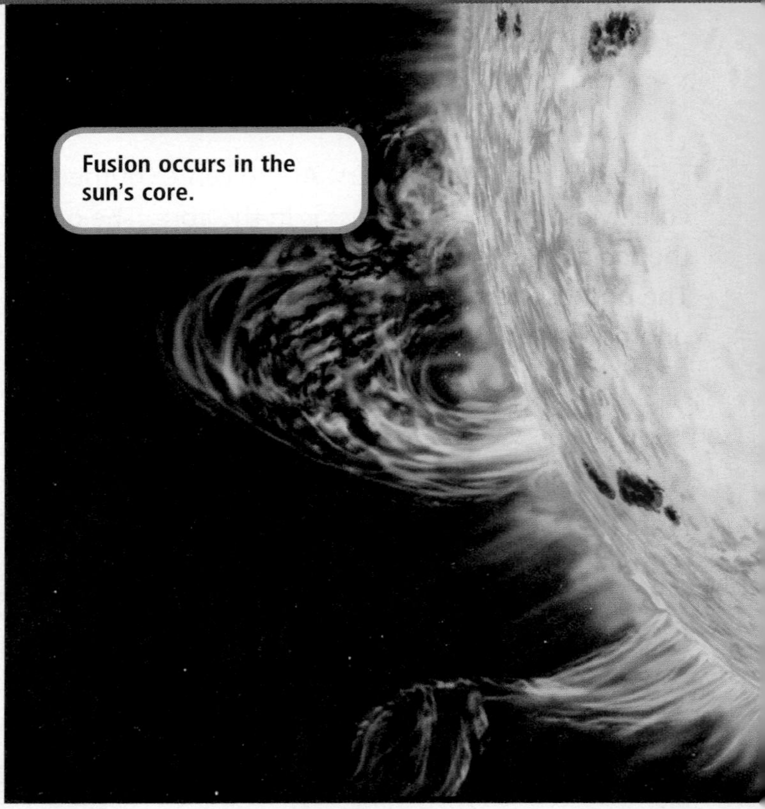
Fusion occurs in the sun's core.

How the Sun Produces Energy

Like other stars, the sun is made up of gases. These gases are mostly hydrogen and helium. The sun's energy comes from **fusion**. Fusion is the joining of small particles into larger ones.

The temperature at the center of the sun is about 15 million °C (27 million °F). Under pressure, hydrogen atoms smash into each other. This produces helium and energy. The sun releases the energy as light and heat.

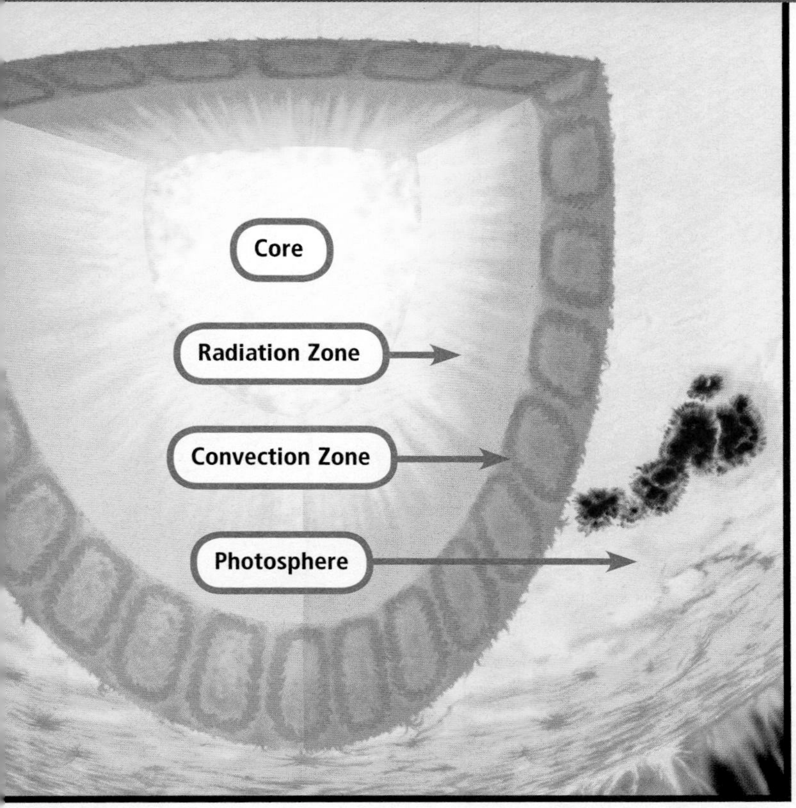
Core
Radiation Zone
Convection Zone
Photosphere

Lesson Review

Complete this Main Idea statement.

1. The ____________ is a star at the center of our solar system.

Complete these Detail statements.

2. The sun is made up of gases, mostly ____________ and ____________.
3. The sun releases ____________ as light and heat.
4. The sun's energy comes from ____________, the combining of small particles into larger ones.

California Standards in This Lesson

5.b *Students know* the solar system includes the planet Earth, the Moon, the Sun, eight other planets and their satellites, and smaller objects, such as asteroids and comets.

Vocabulary Activity

Solar System

There are many parts to the solar system in which we live. In this lesson, you'll learn about some of them.

Fill in the chart to show the number of syllables in each vocabulary term.

Vocabulary Term	Number of Syllables
solar system	
planet	
satellite	
asteroid	
comet	

Lesson 2

What Makes Up the Solar System?

VOCABULARY
solar system
planet
satellite
asteroid
comet

Mercury
Venus
Earth
Mars
Jupiter
Saturn
Uranus
Neptune
Pluto
Sun

Our **solar system** includes nine planets. They revolve around the sun.

Earth is a **planet**. It is a body that revolves around a star.

A **satellite** revolves around Earth. They can be artificial or natural.

An **asteroid** can leave craters when it hits Earth.

A **comet** can have a tail millions of kilometers long.

Hands-On Activity
Model Planets

1. Use clay to model the sun and the planets in the solar system.
2. Place the models on a sheet of paper in order from the sun, and label them.
3. How can a model help you understand the solar system?

____

____

____

✓Concept Check

1. The **Main Idea** on these two pages is There are inner and outer planets. Underline one detail about inner planets and one about outer planets.

2. How is a planet different from a satellite?

____

____

3. Look at the picture. Which planets is Earth near?

____

4. What is one detail about the inner planets?

____

5. Look at the picture. Order the inner planets by size, starting with the smallest.

   1. ____________________
   2. ____________________
   3. ____________________
   4. ____________________

The Inner Planets

A **solar system** is made up of a star and the objects that revolve around it. These objects can be planets, satellites, or other objects. A **planet** is a body that revolves around a star. A **satellite** is a body that revolves around a planet.

Our solar system includes nine planets. There are four inner planets. The inner planets are Mercury, Venus, Earth, and Mars. They are all rocky and dense.

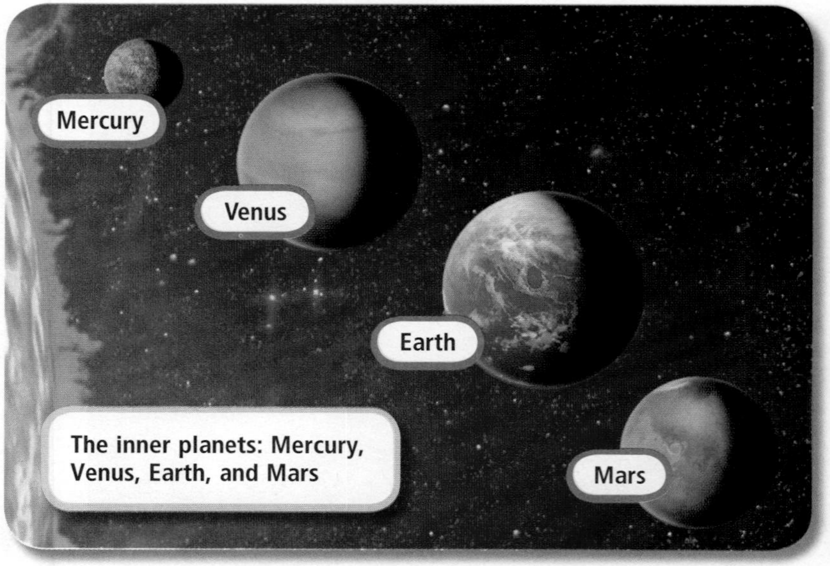
Mercury
Venus
Earth
Mars

The inner planets: Mercury, Venus, Earth, and Mars

The Outer Planets

Beyond the inner planets are the five outer planets. In order from the sun, they are Jupiter, Saturn, Uranus, Neptune, and Pluto.

Jupiter is the largest planet. Saturn is best known for its rings. Uranus has many moons and rings. Neptune has the strongest winds in the solar system. Pluto is small and rocky. In 2006, the International Astronomical Union defined a planet as a body that orbits the sun, is spherical, and is large enough to clear its orbit. They reclassified Pluto as a "dwarf planet," because it is not large enough to clear its orbit.

Pluto
Saturn
Uranus
Jupiter
Neptune

The outer planets: Jupiter, Saturn, Uranus, Neptune, and Pluto

✓ Concept Check

1. What is one detail about the outer planets?

___

2. Fill in the chart. Tell which planet has the feature.

Planet	Feature
	Has the strongest winds in the solar system
	Has many moons and rings
	Best known for its rings
	The largest planet

3. How many outer planets are there?

___

4. List all eight planets in order from the sun.

___

___

✓Concept Check

1. The **Main Idea** on these two pages is Asteroids and comets are objects in space. Underline one detail about asteroids and another about comets.

2. In the space below, draw a picture of two planets to show where the asteroid belt is located. Label the planets and the asteroid belt.

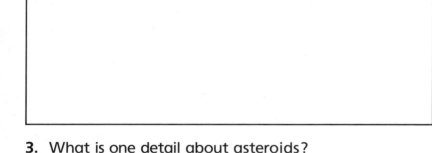

3. What is one detail about asteroids?

____

4. How are asteroids and comets alike?

____

What are comets made of?

Asteroids and Comets

Asteroids are chunks of rock too small to be called planets. Asteroids are located in the area between Mars and Jupiter. They make up the *asteroid belt.*

Meteors are other small pieces of rock that travel through space. When meteors hit Earth's atmosphere, they usually burn up.

A **comet** is a ball of ice, rock, and frozen gases that orbits the sun. A comet may pass close to the sun and then swing out to the edge of the solar system.

Lesson Review

Complete this Main Idea statement.

1. A ____________ ____________ is made up of a star and all the planets and other objects that revolve around that star.

Complete these Detail statements.

2. Our ____________ ____________ includes nine planets.

3. There are ____________ inner planets.

4. There are ____________ outer planets.

California Standards in This Lesson

5.c Students know the path of a planet around the Sun is due to the gravitational attraction between the Sun and the planet.

Vocabulary Activity

Gravity

1. The chart below shows the roots and suffixes of some of your vocabulary words. Use their meanings to tell what your vocabulary words mean.

Vocabulary Word	Parts	Vocabulary Word Meaning
inertia	Root: inert: unable to move Suffix: -ia: state of being	
gravity	Root: gravis: heavy Suffix: -ity: state of	

2. Use the words *orbit* and *elliptical* in a sentence.

____

Lesson 3

What Holds the Moon and Planets in Place?

VOCABULARY
orbit
elliptical
inertia
gravity

The Earth's **orbit** is around the sun.

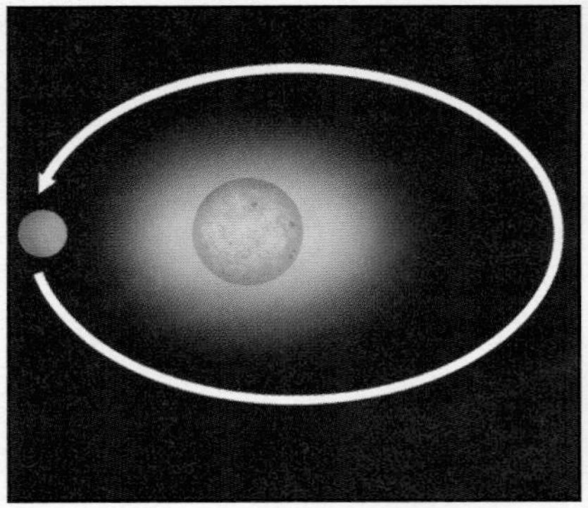

The orbit of Mars is **elliptical**, or oval-shaped.

It takes a lot of force to overcome a rocket's **inertia**.

Gravity helps keep roller coaster cars on the track.

Hands-On Activity
Gravity in Action

Use a ball to watch how gravity and inertia affect objects on Earth's surface. Do this activity inside on an even floor.

1. Hold a ball up above your head and let go. What happens? Why?

____

2. Roll the ball straight ahead of you. What happens to the ball?

____

3. What force do you think made it stop?

____

____

____

Concept Check

1. A **Cause** is something that makes another thing happen. An **Effect** is the thing that happens. Underline the sentence that explains what causes one planet to be closer to the sun than another at certain times. Circle an effect of inertia.

2. What is one effect of an orbit's shape on the planet's distance from the sun?

___

3. Look at the picture. Circle the elliptical orbit.

4. What is an orbit?

___

___

The Path of a Planet Around the Sun

An **orbit** is the path that a body follows as it revolves around another body. The sun is at the center of the solar system. Planets move in **elliptical** orbits around the sun. An elliptical orbit is shaped like a flattened circle. Because the orbits are not circular, each planet is closer to the sun at certain times.

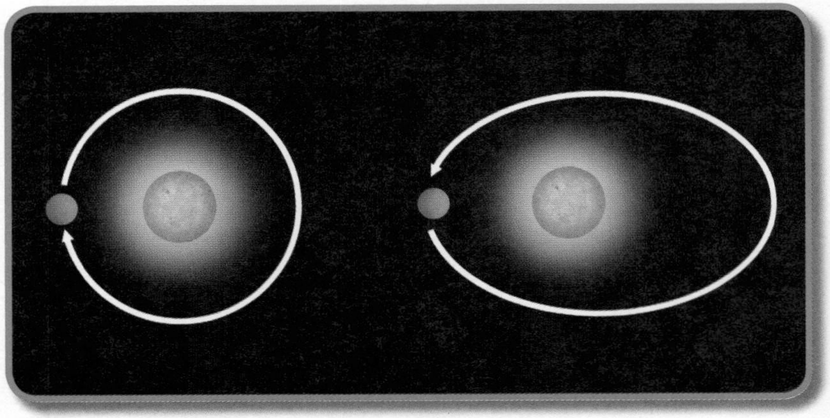

Planets travel around the sun in elliptical orbits.

What Holds the Moon and Planets in Orbit?

What keeps the planets moving in their orbits?

Inertia keeps an object moving in a straight line at a steady speed. The object will keep going unless pushed or pulled by some force.

Gravity is the force that pulls objects toward Earth. If there were no gravity, Earth would travel forever, in a straight line through space. Gravity causes a pull between the mass of a planet and the mass of the sun.

Gravitational force plus inertia produces a planet's orbit. The sun's constant pull changes a planet's direction. It causes the paths of planets and the moon to be curved.

The Earth and the Moon
Inertia and gravity keep the moon in a nearly circular orbit around Earth.
Earth and its moon move together around the sun. Because the moon is always orbiting Earth while the Earth-moon system orbits the sun, the moon's path around the sun is a series of loops.

✓ Concept Check

1. Read the caption by the picture. In the space below, draw the moon's orbit around the sun.

2. What causes the planets to stay in their orbits?

____

3. Why do the planets orbit the sun?

____

____

4. What keeps an object in a straight line unless it is pushed or pulled?

____

✓Concept Check

1. A **Cause** is something that makes another thing happen. An **Effect** is the thing that happens. Underline the sentence that explains what causes the moon to look different every day. Circle an effect of the moon's being between the Earth and the sun.

2. What is meant by the moon's *phase*?

____

____

3. Between a new moon and a full moon, do you see MORE of the moon each night, or LESS?

____

4. Label each phase of the moon shown below.

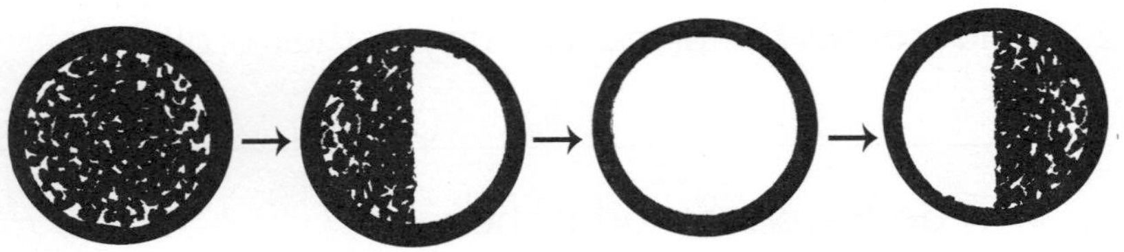

__________ __________ __________ __________

Phases of the Moon

The way the moon looks from Earth changes every day. Half of the moon is always lit by the sun. How much you see depends on the moon's phase. A moon *phase* is the shape the moon seems to have as it orbits Earth.

When Earth is between the moon and the sun, you see a full moon. When the moon is between Earth and the sun, you can't see the moon. This is the new moon. Starting with the new moon, you see more and more of the moon each day until you see the full moon.

waning gibbous
third quarter
waning crescent

What are the phases of the moon?

full moon
waxing gibbous
second quarter
first quarter
new moon

✓Concept Check

1. Look at the picture below. What will the moon look like from Earth?

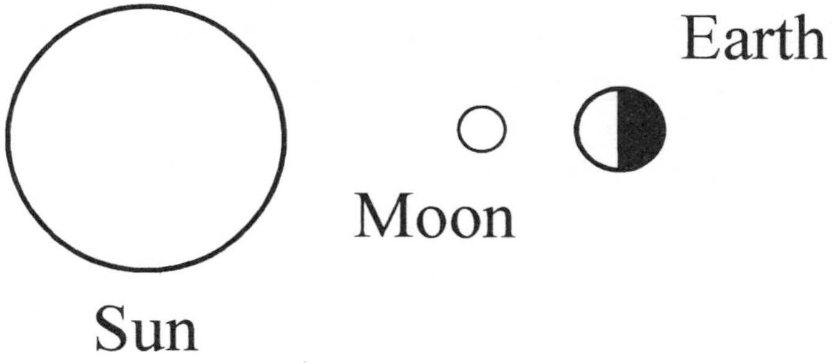
Earth
Moon
Sun

2. Draw a picture that shows four phases of the moon. Label the phases.

✓Concept Check

1. A **Cause** is something that makes another thing happen. An **Effect** is the thing that happens. Underline the sentence that explains what causes space shuttles to orbit Earth instead of the sun. Circle the effect of inertia on a space shuttle.

2. What causes astronauts to appear weightless in space?

____

3. What effect does gravity have on the space shuttle?

____

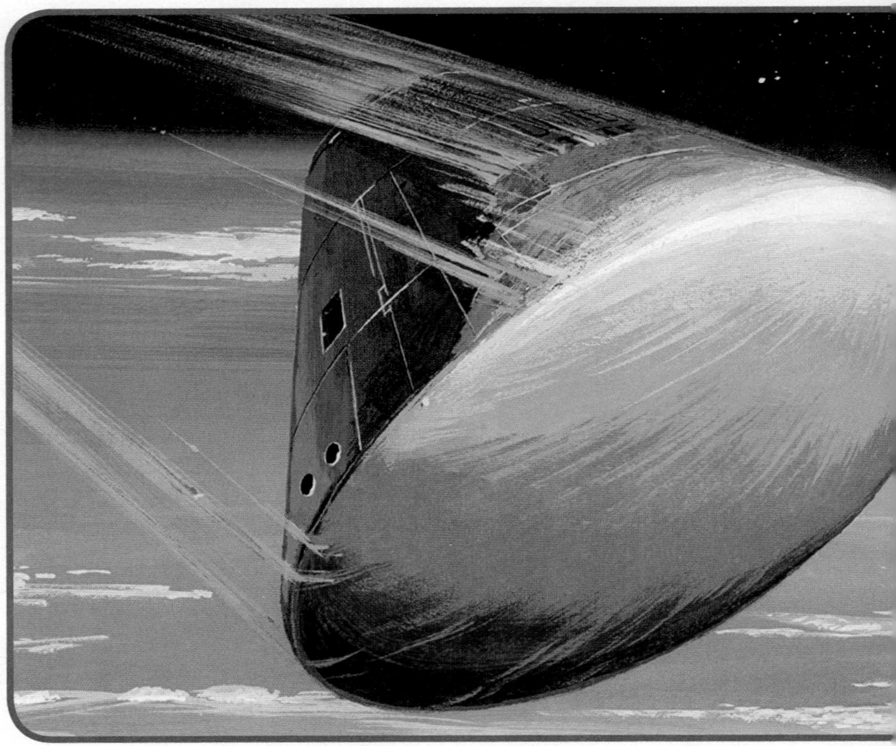

Gravity in Space

Astronauts in the space shuttle orbit close to Earth. The shuttle is so close that gravity is almost as strong there as it is on Earth. Astronauts appear to be weightless because inertia and gravity are balanced.

The shuttle engines are off but the shuttle moves forward. That is inertia. Earth's gravity pulls on the shuttle as it moves forward. The astronauts inside are really free-falling as the shuttle moves forward.

Lesson Review

Complete these Cause and Effect statements.

1. The ___________ of a planet is due to gravitational attraction between the sun and the planet.
2. ___________ and ___________ cause planets to stay in their orbits.
3. How much of the moon you see depends on the moon's ___________.
4. Astronauts appear to be weightless because inertia and gravity are ___________.

Unit 5 Review

Circle the letter in front of the best choice.

1. Where is the largest body in the solar system located?

   A on Earth
   B at the center of the solar system
   C at the outer edge of the solar system
   D Earth is the largest body.

2. How does the sun compare to other stars?

   A It is bigger.
   B It is smaller.
   C It is hotter.
   D It is average.

3. What gases is the sun mostly made of?

   A hydrogen and oxygen
   B helium and argon
   C hydrogen and helium
   D oxygen and helium

4. What is fusion?

   A the joining of small particles into larger ones
   B the breaking of larger particles into smaller ones
   C a form of waves
   D a layer of the sun

5. Which is NOT an inner planet?

   A Mars
   B Venus
   C Earth
   D Saturn

6. Which is NOT an outer planet?

   A Jupiter
   B Uranus
   C Mercury
   D Neptune

7. Which object does NOT orbit another in space?

   A planets
   B meteors
   C asteroids
   D comets

8. What do planets orbit?

   A stars
   B the moon
   C comets
   D other planets

9. What shape are the planet's orbits around the sun?

   A They are all different.
   B They are all circular.
   C They are all elliptical.
   D They all move in a straight line.

10. There is a new moon out. What phase will you see NEXT?

A full moon
B first quarter
C waning crescent
D waxing gibbous

11. Which is TRUE of the moon?

A It is always entirely lit by the sun.
B The sun never lights it.
C Half of the moon is always lit by the sun.
D Starting with a full moon, you see more of the moon each night.

12. A body in space that orbits a larger body is a

____

13. The sun produces energy by fusion. What happens during fusion?

____

____

____

____

____

14. Which two forces keep planets in an orbit around the sun?

____

____

____

15. Look back to the question you wrote on page 156. Do you have an answer for your question? What did you learn that helps you understand the solar system?

____

____

____

____

____

____

____

____

____

____

____

____